PATENTLY Ridiculous

PATENTLY Ridiculous

FIG. 1

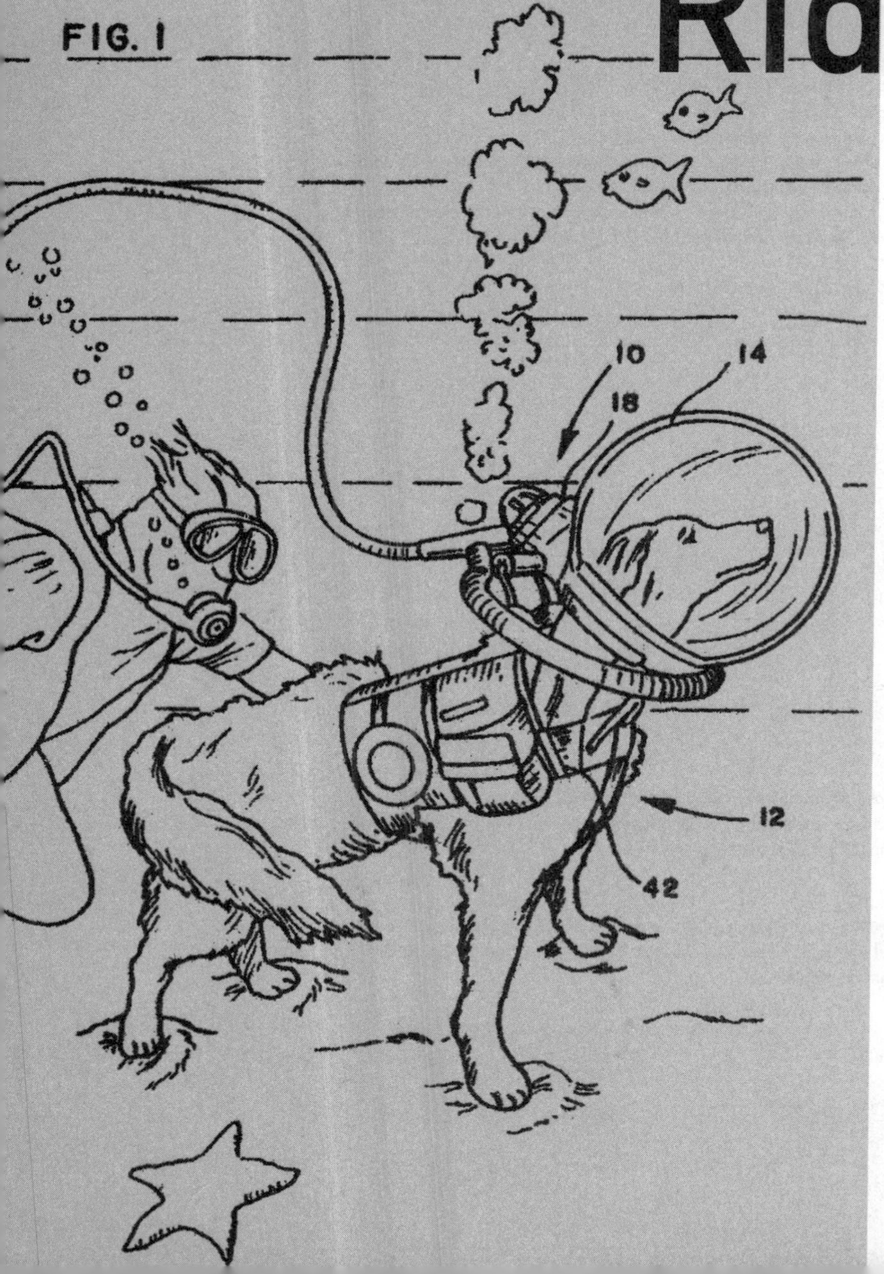

10
18
14
12
42

scuba-diving dogs,
beerbrellas,
musical toothpaste,
and other
patented strokes
of genius

Richard Ross

A PLUME BOOK

PLUME
Published by the Penguin Group
Penguin Group (USA) Inc., 375 Hudson Street, New York, New York 10014, U.S.A.
Penguin Group (Canada), 10 Alcorn Avenue, Toronto, Ontario, Canada M4V 3B2 (a division of Pearson Penguin Canada Inc.)
Penguin Books Ltd, 80 Strand, London WC2R 0RL, England
Penguin Ireland, 25 St Stephen's Green, Dublin 2, Ireland (a division of Penguin Books Ltd)
Penguin Group (Australia), 250 Camberwell Road, Camberwell, Victoria 3124, Australia (a division of Pearson Australia Group Pty Ltd)
Penguin Books India Pvt Ltd, 11 Community Centre, Panchsheel Park, New Delhi—110 017, India
Penguin Books (NZ), Cnr Airborne and Rosedale Roads, Albany, Auckland, New Zealand (a division of Pearson New Zealand Ltd)
Penguin Books (South Africa) (Pty) Ltd, 24 Sturdee Avenue, Rosebank, Johannesburg 2196, South Africa

Penguin Books Ltd, Registered Offices: 80 Strand, London WC2R 0RL, England

First published by Plume, a member of Penguin Group (USA) Inc.

First Printing, April 2005
10 9 8 7 6 5 4 3 2 1

CIP data is available.
ISBN 0-452-28587-9

Printed in the United States of America
Set in DIN Mittelschrift
Designed by Laura Lindgren

contents

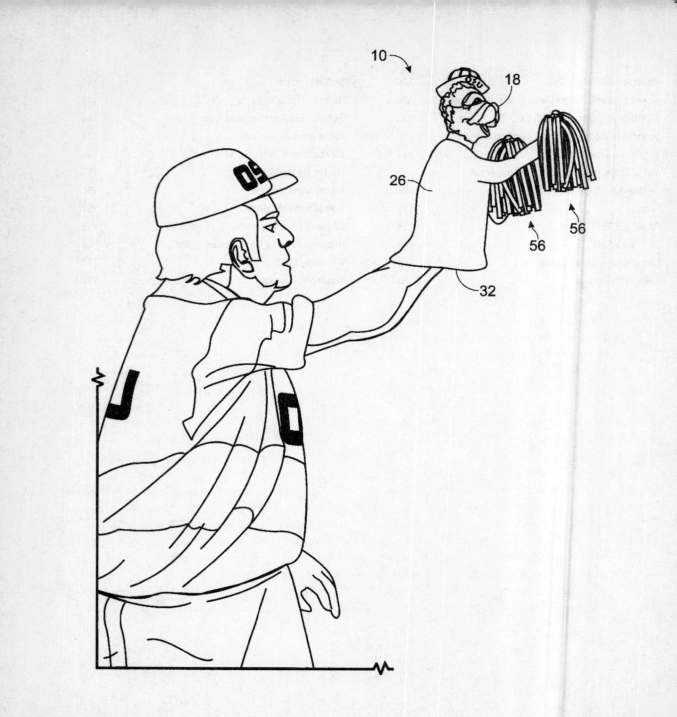

foreword

Jonas Salk was a brilliant scientist. His creation of a vaccine to prevent polio was a blessing to the world. Orville and Wilbur combined methodical engineering and inspired thinking to create powered flight. Dr. John S. Pemberton originated the Coca-Cola formula. Milestones all, but what about the *other* guys, those with more modest aspirations, like wipers for headlights? Could the Wright brothers have imagined they were in fact creating the opportunity for Robert Plath to invent the rolling carry-on?

Heroes whose patented inventions have changed the world for millions of people fill the pages of countless biographies; their lives are defined by genius, daring, and hard work. But what about the *other* guys and *beerbrellas*? Does the inventor of the dual nose picker feel this is his legacy, his contribution to humanity?

Many, if not most, inventions are the fruit of relentless effort and careful methodology. This is the scientific method: two steps forward, one back—experiment and control, aspiration and perspiration. And then there are the *other* guys and the wobbling headpiece.

James Wright, a GE engineer, came upon Silly Putty by mixing silicone oil with boric acid. Alhazen (Ibn Al-Haytham) invented the first pinhole camera in approximately A.D. 1000. Louis-Jacques-Mandé Daguerre made the first practical photographic image some 170 years past. These are inventions of varying degrees of significance, but who could challenge as top dog the Armenian inventor Spandjian, who created one of the greatest inventions of the modern world, spandex. George Bernard Shaw called condoms "the greatest invention of the 19th century," although he might not have looked too deeply into the pyramids of Egypt some 3,000 years prior (where condoms were found—unpatented), but perhaps he was referring to Charles Goodyear and the invention of rubber.

Ermal Fraze invented the pop-top. Hinda Miller and Lisa Lindahl invented the sports bra. These may be the middle ground for practical applications that have become ubiquitous in modern society. But there are always the extremes. For all the teams of scientists working in laboratories to cure cancers, transplant organs, and make nonfossil fuels, there are thousands of guys and women sitting in garages working on a better #1 foam finger, an improved golf ball washer, synthetic hair plugs, or an updated Flowbie.

Simple ideas that represent more than ingenuity, they express faith and entrepreneurship at a high level—well, maybe these are more about faith.

After all, a scuba suit for your golden retriever? Is the underwater pooper scooper far behind?

Patent No. US 6,663,462

Dec 16, 2003

Thomas A. Bettendorf; Michele M. Bettendorf;

Daryl D. Lovejoy; Louis C. Lovejoy

Aggression-Relieving Stuffed Doll

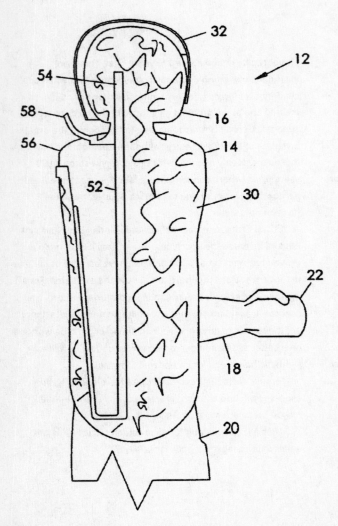

Abstract

An aggression-relieving stuffed doll includes a doll body having a torso, arms, and legs, the doll body being formed of a soft material and filled with soft, stuffing material. A rigid plate is situated in the torso of the doll body and is substantially impenetrable. The invention includes a plurality of pins, each pin having a blunt end simulative of a sports ball and a sharp end capable of penetrating the doll body but not the inner torso plate. Therefore, a user may stick pins into the doll to relieve aggression caused by the simulated sports player, although the pin is prevented by the inner plate from penetrating through the back of the doll and into the user's hand.

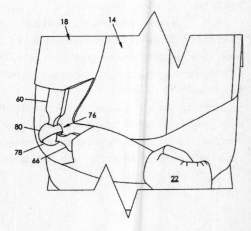

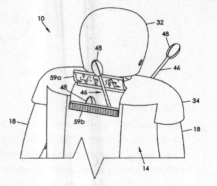

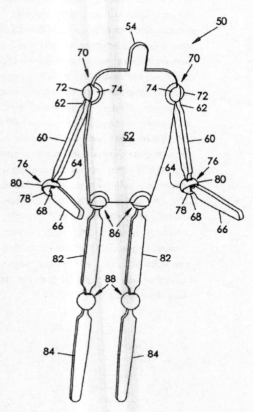

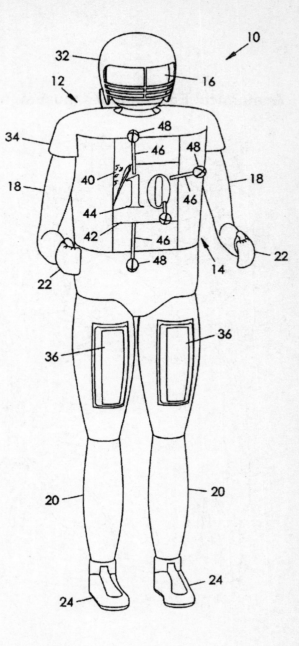

Patent No. 4,865,550
Sept 12, 1989
Shao-Chun Chu

Anatomical Educational Amusement Ride

Fig.1

Abstract

An educational amusement apparatus forms a large building structure having an external appearance simulating a man and a woman resting partially under a blanket, wherein riders are taken through a succession of cavities that simulate internal organs of the man and woman. Entrance to a head chamber simulating an oral cavity is achieved by a stairway supported by a simulated arm of the man, the oral cavity having displays of teeth in normal and abnormal conditions, and serving as a staging area for a train to carry the riders. The train passes into a simulated cranial cavity of the woman, past a sectional display of simulated ear organs, and into a body portion of the building that is representative of the abdomen of both the man and the woman, first through a simulated esophagus, stomach, and intestine of an alimentary canal, through simulated urinary and reproductive tracts, then through a simulated liver and a simulated cardiovascular canal, and finally through a simulated lung and windpipe to an exit staging area of the building.

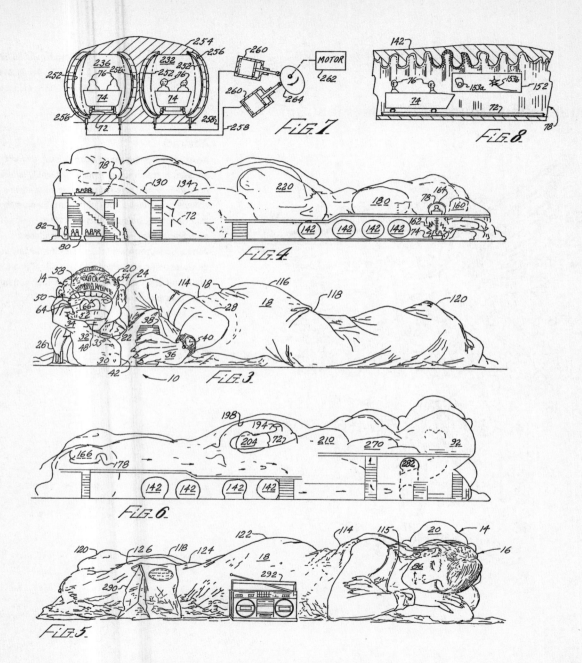

13

Angel Action Figure Doll

Patent No. 5,588,895
Dec 31, 1996
Diana A. Larson

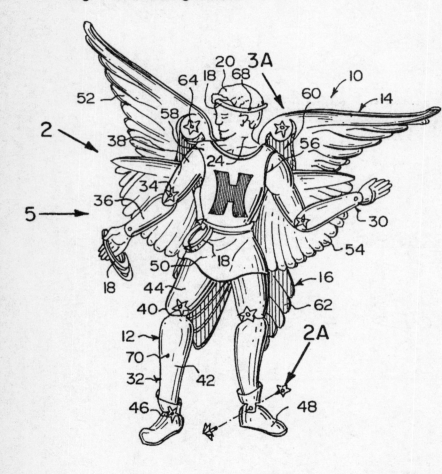

Abstract

An angel action figure doll comprising a small adjustable mannequin. A wing assembly is also provided. A structure is for mounting the wing assembly against a back of the small adjustable mannequin. A halo is worn on a head of a small adjustable mannequin, to simulate a conventional representation of an immortal, spiritual being superior to humans, that is an attendant and messenger of God, which can be used as a plaything of a child.

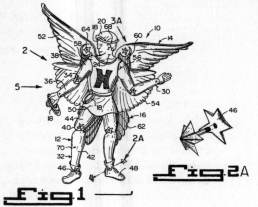

Fig.1

Fig.2A

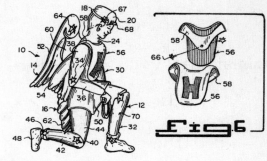

Fig.5

Fig.6

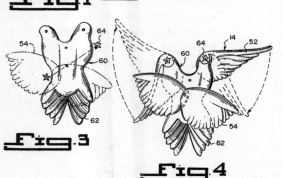

Fig.3

Fig.4

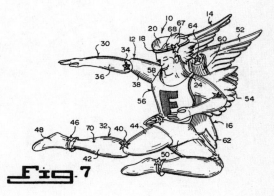

Fig.7

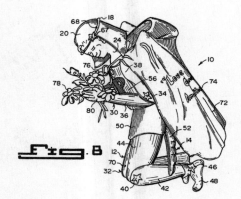

Fig.8

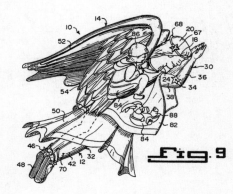

Fig.9

Patent No. 4,778,172
Oct 18, 1988
William C. Bryan

Animal Head

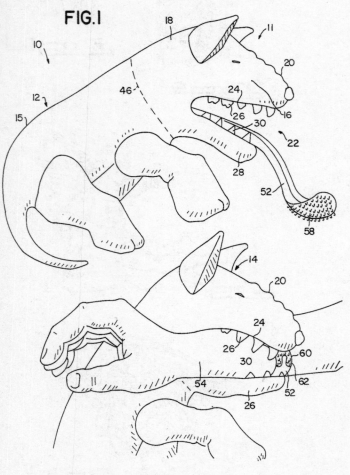

FIG.1

FIG.2

Abstract

An animal head suitable for attachment to a stuffed animal body is molded to form a thin shell and then filled with conforming foam rubber. The head defines open jaws that can fit around a wrist of a person and an elongated tongue that can be wrapped around the wrist and secured between the upper and lower jaws to serve as a carrying handle. By making the tongue flesh tone in color and providing a red coloration on a visible portion of the tongue, a novelty effect simulating blood dripping from a bite is achieved.

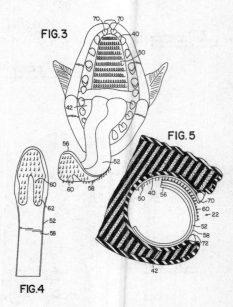

FIG.3

FIG.5

FIG.4

Patent No. 3,017,193

Jan 16, 1962

Oscar R. Klein

Animal Simulating Bicycle

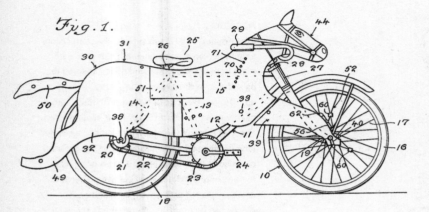

Fig. 1.

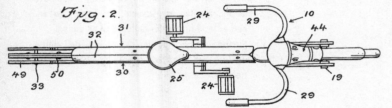

Fig. 2.

Abstract

This invention relates to a bicycle, and more particularly to an animal simulating mechanism for attachment to a bicycle. The object of the invention is to provide a means for converting an ordinary bicycle into a device which simulates an animal such as a horse, so that a child using the device will derive increased pleasure or amusement therefrom. Another object of the invention is to provide a means for providing sounds as the bicycle travels so that the sounds will resemble the sounds produced by a horse's hoofs as the horse moves along an area.

Patent No. 6,090,420

July 18, 2000

Thomas J. Coleman; William K. Schlotter IV;

Princess Ann Coleman; Ann M. Schlotter

Animated Chicken Candy Pop Combination

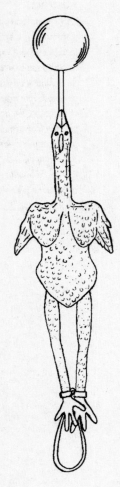

Abstract

An animated body candy pop device comprising a main housing, an expandable sleeve in said main body, a candy stick and candy adapted to be secured to the expandable sleeve. The main housing can be solid or hollow and made with a pliable material to provide a "limp" feeling effect. The lower end of the main housing has a hanger loop for displaying the product. The hanger loop can also fit around a person's wrist or be used to hang onto a belt for easy carrying. The candy can be replaced with any type of lollipop. The device can be made in different body styles such as animals, fish and insects, etc., to provide fun and entertainment for children of all ages. Further, the neck portion can be provided with a noisemaker which will make a noise when the neck is bent.

Patent No. US D447,720
Sept 11, 2001
Michael G. Beller

Animated Golf Bag Novelty

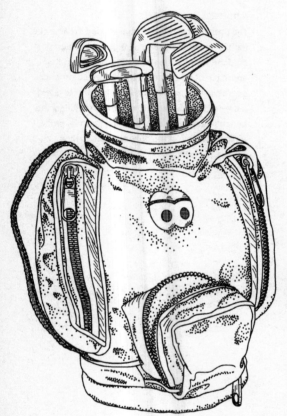

Abstract
A front perspective view of an animated golf bag novelty showing my new design. Also a side perspective view thereof, a side perspective view of the opposite side thereof, a rear perspective view thereof, and a bottom view thereof.

FIG. 2

FIG. 3

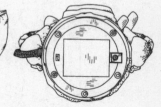

FIG. 4

Patent No. 5,322,717
June 21,1994
William R. Killian

Animated Outdoor Ornament

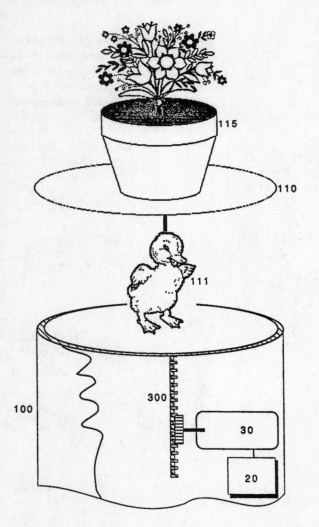

Abstract

The present invention is an animated ornamental fixture for installation in a garden or similar outdoor area frequented by visitors. Upon the approach of an observer, the fixture commences operation to surprise and entertain by elevating a decorative flowerpot to reveal a small figurine beneath, and optionally emitting sounds, and after departure of the observer, the fixture ceases operation.

Patent No. 4,364,132
Dec 21, 1982
Lawrence D. Robinson

Aquarium Bath

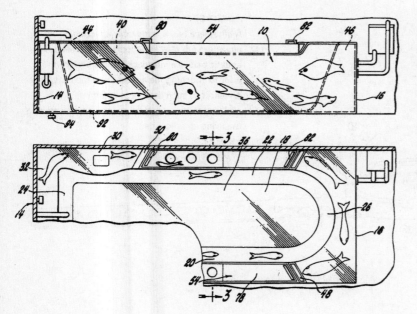

Abstract

A bath tub having inner and outer front, rear and side walls, a ledge extending along the tops of these walls, a base uniting the inner and outer walls, and the whole forming a water-tight chamber extending around the tub. A cover is provided in a ledge of the tub. The cover provides an access opening, and there is also an aeration opening, for admission of air to water in the chamber and for servicing of the chamber. The chamber can be used for fish to thus provide an aquarium-tub combination.

Patent No. US 6,416,217
July 9, 2002
Harold Von Braunhut

Aquarium Watch

Abstract

A timepiece has a removably attachable aquarium adapted to support aquatic life. Living aquatic pets are introduced to the aquarium prior to attaching the aquarium to the timepiece. A wearer of such timepiece is then able to contemporaneously tell time and enjoy watching the living aquatic pets.

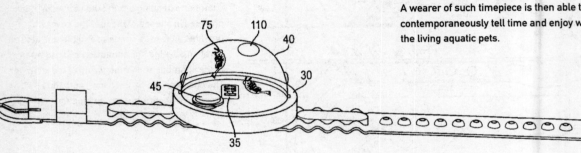

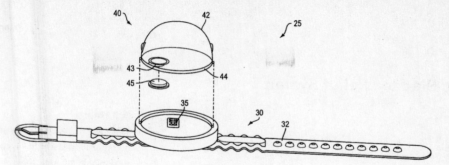

FIG. 1

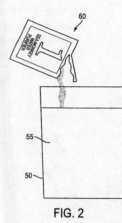

FIG. 2

FIG. 3

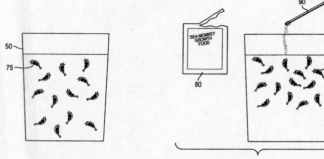

FIG. 4

FIG. 5A

Patent No. US 6,706,027
Mar 16, 2004
Mark R. Harvie

Automatic Bladder Relief System

Abstract

This invention relates to an automatic or semi-automatic bladder relief system, specifically designed to increase the overall sanitation and comfort for users that may require a means to dispose of their urine in the absence of other sanitary facilities due to situations such as aircraft pilots and incontinent individuals.

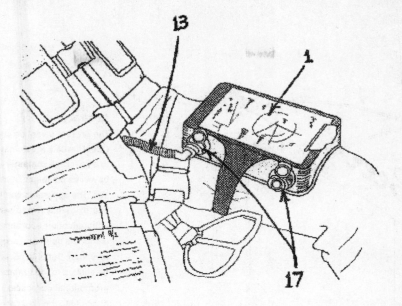

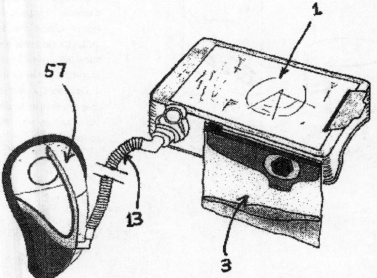

Patent No. US 6,637,447
Oct 28, 2003
Mason Schott McMullin; Robert Platt Bell;
Mark Andrew See

Beerbrella

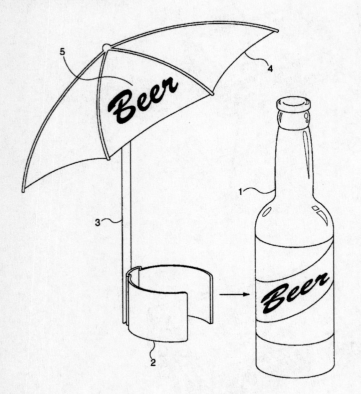

Abstract

The present invention provides a small umbrella ("Beerbrella") which may be removably attached to a beverage container in order to shade the beverage container from the direct rays of the sun. The apparatus comprises a small umbrella approximately five to seven inches in diameter, although other appropriate sizes may be used within the spirit and scope of the present invention. Suitable advertising and/or logos may be applied to the umbrella surface for promotional purposes. The umbrella may be attached to the beverage container by any one of a number of means, including a clip, strap, cup, foam insulator, or as a coaster or the like. The umbrella shaft may be provided with a pivot to allow the umbrella to be suitably angled to shield the sun or for aesthetic purposes. In one embodiment, a pivot joint and counterweight may be provided to allow the umbrella to pivot out of the way when the user drinks from the container.

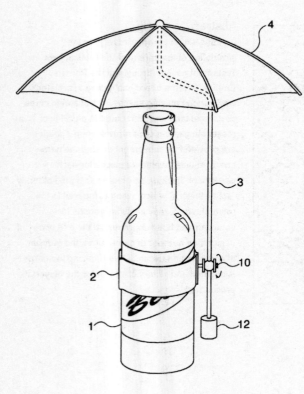

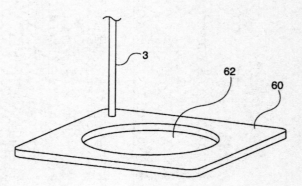

27

Patent No. 4,441,718

April 10, 1984

Mark J. Olson

Biblical Game

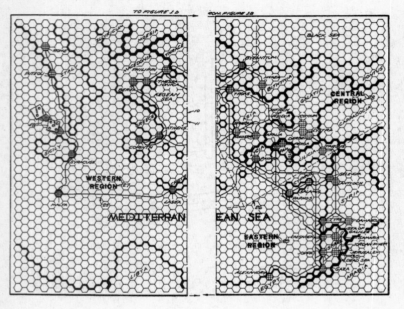

Abstract

An educational game primarily for the simulation of the spread of the Biblical New Testament gospel throughout the Roman Empire. A game board has indicia associated with various playing spaces, representing cities, ports, and land and sea routes. A plurality of moveable game pieces representing Apostles are provided for movement around the game board, at least two game pieces for each participant. Each game piece is assigned other "ability factors" which, when compared to the roll of the dice, may allow an Apostle to establish and build churches. At the end of a designated number of game turns, the number of converts and deacons "won" during the game is counted and victory or defeat for the players is proclaimed.

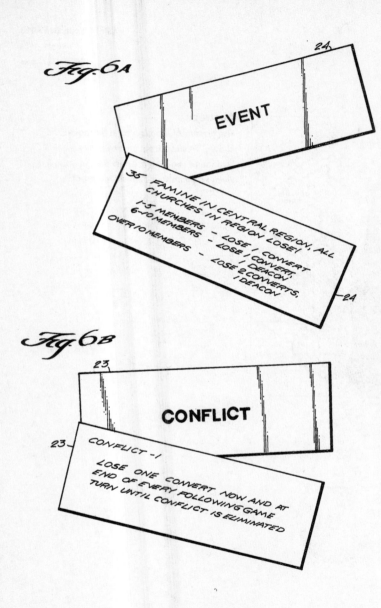

Fig. 6A

EVENT

35 — FAMINE IN CENTRAL REGION, ALL CHURCHES IN REGION LOSE!
1-5 MEMBERS - LOSE 1 CONVERT
6-10 MEMBERS - LOSE 1 CONVERT, 1 DEACON
OVER 10 MEMBERS - LOSE 2 CONVERTS, 1 DEACON

24

Fig. 6B

CONFLICT

23 — CONFLICT - 1
LOSE ONE CONVERT NOW AND AT END OF EVERY FOLLOWING GAME TURN UNTIL CONFLICT IS ELIMINATED

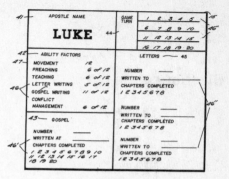

APOSTLE NAME		GAME TURN	1 2 3 4 5
LUKE 44			6 7 8 9 10
			11 12 13 14 15
			16 17 18 19 20

ABILITY FACTORS		LETTERS — 45
MOVEMENT	12	NUMBER ____
PREACHING	6 of 12	WRITTEN TO
TEACHING	6 of 12	CHAPTERS COMPLETED
LETTER WRITING	5 of 12	1 2 3 4 5 6 7 8
GOSPEL WRITING	11 of 12	
CONFLICT MANAGEMENT	6 of 12	NUMBER ____ WRITTEN TO CHAPTERS COMPLETED 1 2 3 4 5 6 7 8

43 — GOSPEL

NUMBER ____
WRITTEN AT
CHAPTERS COMPLETED
1 2 3 4 5 6 7 8 9 10
11 12 13 14 15 16 17
18 19 20

NUMBER ____
WRITTEN TO
CHAPTERS COMPLETED
1 2 3 4 5 6 7 8

Fig. 2

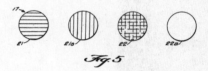

Fig. 5

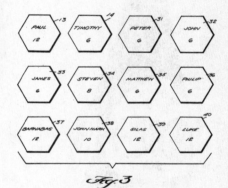

Fig. 3

Fig. 4

Patent No. US D173,440

Nov 9, 1954

William P. Kupka

Birdhouse

Abstract

The ornamental design for a birdhouse, as
shown. This is a front perspective view of a
birdhouse showing my new design. Also a rear
view, and a bottom plan view thereof.

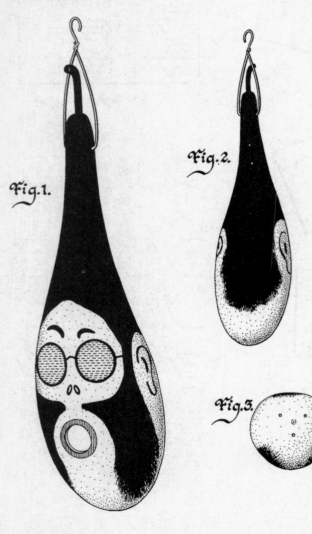

Fig.1.

Fig.2.

Fig.3.

Body Dryer with Mirror

Patent No. US 6,349,484

Feb 26, 2002

Sol Cohen

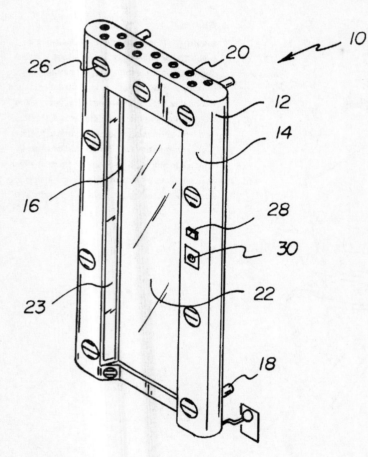

Abstract

A body dryer is provided including a housing having an elongated height. Also included is a mirror mounted on the housing. A heating mechanism is situated along the height of the housing for expelling heated air therefrom.

Patent No. 4,241,721

Dec 30, 1980

Gordon L. Holly

Body Warmer

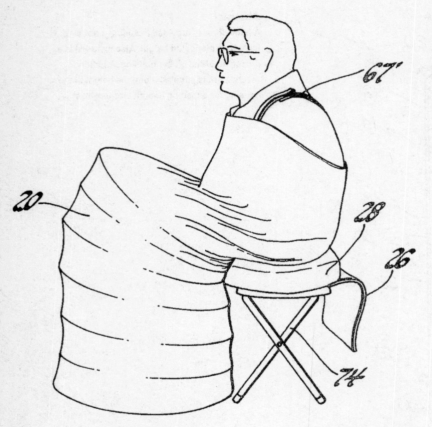

Abstract

A body warmer is provided that includes a platform, a heat source, an envelope surrounding the user, and hoops in the envelope. The envelope is kept from contacting the heat source by the hoops. The user is kept from contacting the heat source by a shield. The heat source is pivotally connected to the platform and movable from a first position in which it is stored to a second position in which it is used. A cushion is attached to the inside of a cover that seals the body warmer for storage.

Patent No. US 6,193,578

Feb 27, 2001

Thomas Carl Weber

Bubbling Brain Novelty

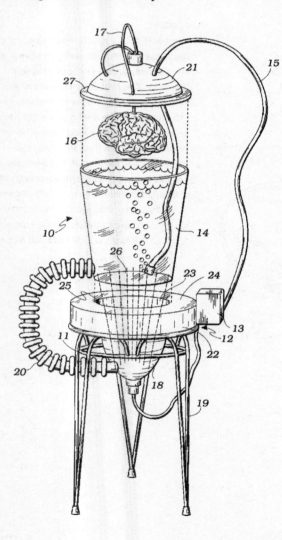

Abstract

A transparent vessel containing a fluid and a life-like full scale human brain inserted into the fluid is used for novelty purposes. The vessel is open on its top and sealed on its bottom, and is mounted on a base portion of the device. Air bubbles are produced from an air pump attached to the side of the base portion. A tube connected to the air pump and an outlet port is positioned inside of the vessel, immersed in the liquid to supply air. The transparent, water-tight brain vessel is lit from underneath by a lamp. To enhance an effect of a scientific fiction experiment, the novelty is placed on a stand and has one or more decorative perforated tubes attached for visual effect. The tank is preferably covered with a dome-shaped lid.

Patent No. US 6,206,000
Mar 27, 2001
Dwane L. Folsom

Canine Scuba Diving Apparatus

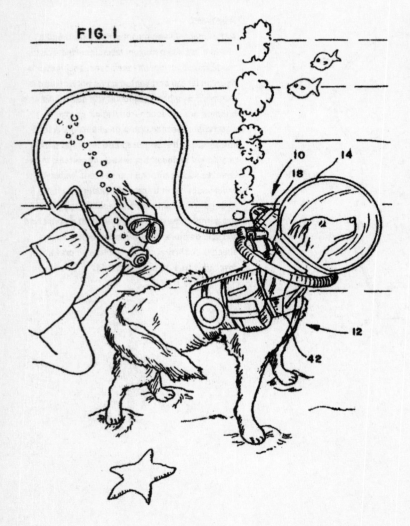

FIG. 1

Abstract

The invention is a special modified scuba diving apparatus intended for use by an animal, and more specifically the famous diving dog "Shadow." The invention includes a helmet, a harness for supporting the helmet and a source of breathable gas to the interior of the helmet, an exhaust for withdrawing exhaled air and residue water from the helmet without depressurizing the helmet, and a system of weights to compensate for the buoyancy of the user, and to counteract a net moment created about the center of buoyancy. The breathing system includes a muffler. An intercom system for voice instructions to Shadow can also be included.

Patent No. US 6,641,403
Nov 4, 2003
Janet R. Bavasso

Child Shoe-lacing and Dressing Learning Kit

Abstract

A child shoe-lacing and dressing learning kit for teaching a child between the ages of 3 and 6 to tie shoes and dress oneself. The child shoe-lacing and dressing learning kit includes a pair of strap members each having an outer side and first and second longitudinal edges and being adapted to fasten about thighs of seated users; and also includes a pair of shoe members being removably fastened upon the outer sides of the strap members and having shoe lace holes therein; and further includes shoe laces for lacing up the shoe members; and also includes fastening members being attached to the strap members and to the shoe members.

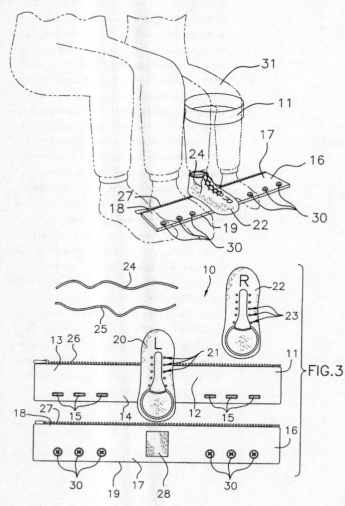

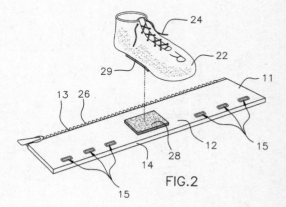

FIG.3

FIG.2

Patent No. US 6,397,389
June 4, 2002
Nils C. Schultz

Child Walking Harness

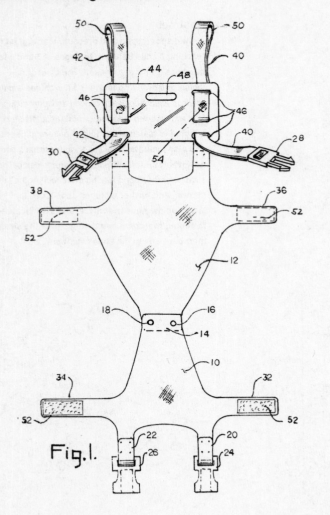

Fig.I.

Abstract

A child walking harness made of flexible material comprising a front and rear body panel which are connected through the legs of the child at the crotch portion. The front and rear body panels are provided panel extensions that connect to each other by means of hook and loop fasteners attached to each end thereof. The upper portion of the body panels are provided with shoulder straps which are fitted with releasable securing devices which lock together over the child's shoulder. The rear body panel straps are looped together and fed through a flexible rear adjustment plate which allows the two upper straps to make adjustments to the size of the child and forms a hand-hold at the upper end to support the child while the child is balancing itself to walk and give comfort and safety to the child and relief of back stress to the parent. The adjustment plate provides a separation of the holding straps to prevent closure of the straps around the child's head and neck. The adjustment plate is also provided with an aperture to allow a leash to be attached to it to control the movement of the child after the child is more advanced in walking.

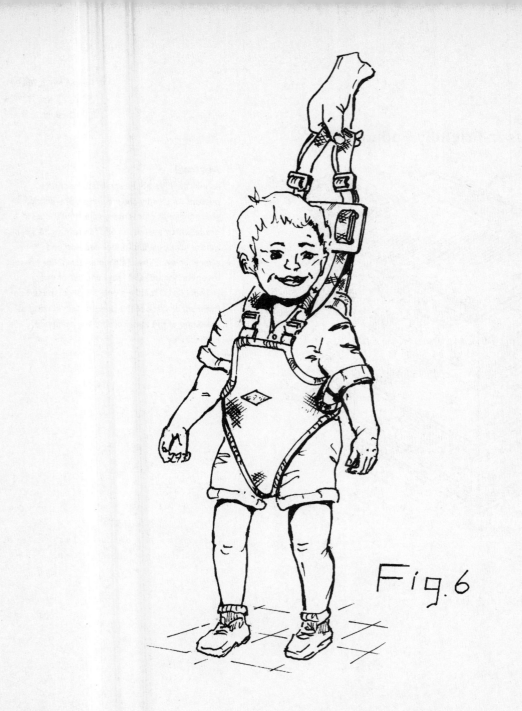

Fig. 6

Patent No. 4,784,382

Nov 15, 1988

Elizabeth A. Myers

Children's User-Friendly Podium

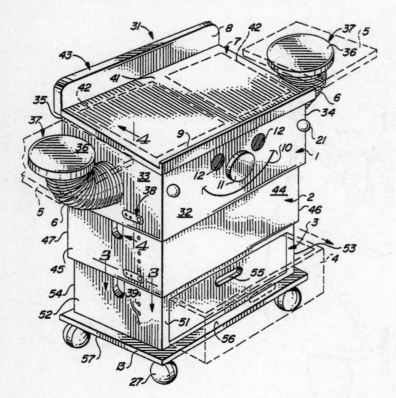

Abstract

A height-adjustable user-friendly, child's podium for use by children from pre-school years through elementary school to the sixth grade which assists a child in learning to speak before groups without fear, tension and apprehension. The child's podium of the present invention includes at least one set of user-friendly facial features operatively disposed on the front or back of the podium for relaxing at least one of the speaker and the audience, respectively.

Patent No. 3,464,151
Sept 2, 1969
Robert L. Motley

Child's Rattle with Bells and Simulated Animal

Abstract

A child's rattle having a hollow body and a hollow handle extending therefrom, each having sound producing elements therein. An animal is universally supported in the body which has transparent characteristics to enable observation thereof with the animal including a movable component operative in response to movement of the rattle.

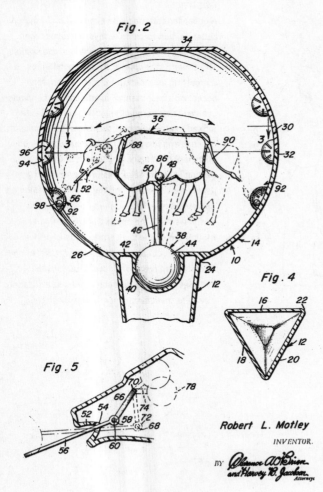

Fig. 2

Fig. 4

Fig. 5

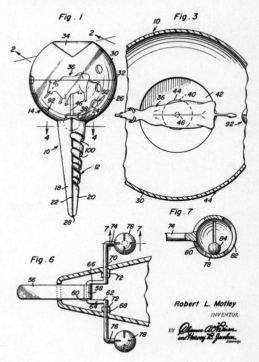

Fig. 1

Fig. 3

Fig. 7

Fig. 6

Robert L. Motley
INVENTOR.

BY

Robert L. Motley
INVENTOR.

BY

Patent No. US 6,490,999 B1

Dec 10, 2002

Donald Robert Martin Boys

Collar Apparatus Enabling Secure Handling of Snake by Tether

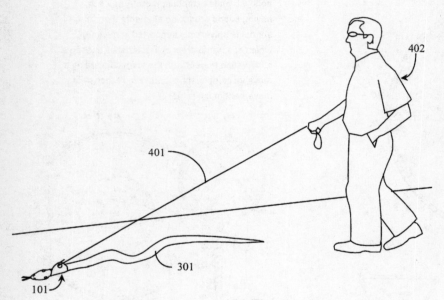

Fig. 4

Abstract

A collar for collaring a snake has an elongated collar section forming a physical collar when wrapped around the body portion of the snake. The collar further has a support section for supporting an attachment mechanism for accepting attachment of a tether and a connector system comprising at least two components affixed to strategic portions of the collar section for securing the collar in place around the body portion of the snake. The length of the collar section is such that a portion thereof overlaps itself when fitted around the snake providing an adjustable interface containing separate components of the connector system whereby mating the connector components together secures the collar in place on the snake. In one embodiment the collar apparatus further includes a concertina movement-neutralization device for reducing concertina movement through the collar.

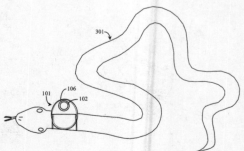

Patent No. US 6,327,997 B1

Dec 11, 2001

Olivia A. Terry, Lauren O. Dipolito

Combination Aquarium and Furniture System

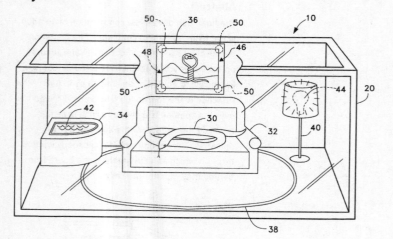

Abstract

A combination aquarium and furniture system constructed to accommodate small animals in a human-like setting and for enhancing the aesthetic nature of various living environments, which includes both saltwater and freshwater. To complement the system, a backdrop including a panoramic scene is optionally secured onto an aquarium. The furniture and related items provide an environment which includes a resting place, drinking source, illumination and/or heat source, and an oxygen generating device.

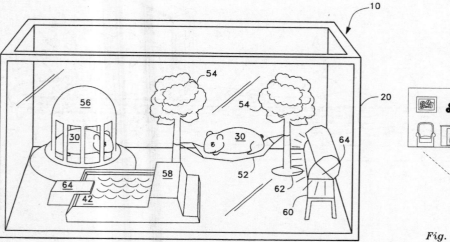

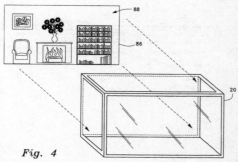

Fig. 4

Patent No. 5,766,052
June 16, 1998
Martin C. Metro; Naomi R. Fine

Combination Child Float/Adult Aquatic Exercise Device

Abstract

The floatation device is a combination of an adult water exerciser and a child float. The floatation device allows a child to float safely and securely on the water while an adult uses the device as a kickboard to push it or as a leg buoy to pull it. Attached to the rear of the board and directed over the opening is a canopy which is designed to shield the child from the sun and from splashes. The canopy includes a canopy window through which an adult pushing the board from behind may monitor the child.

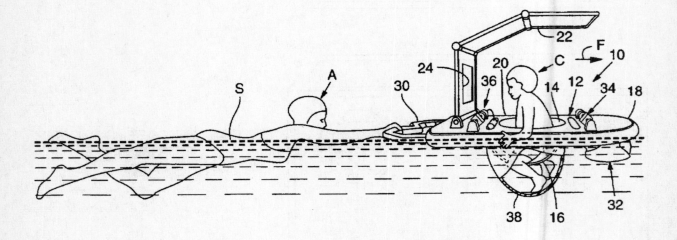

FIG. 1

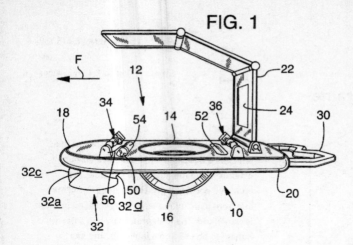

FIG. 2

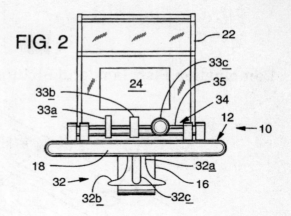

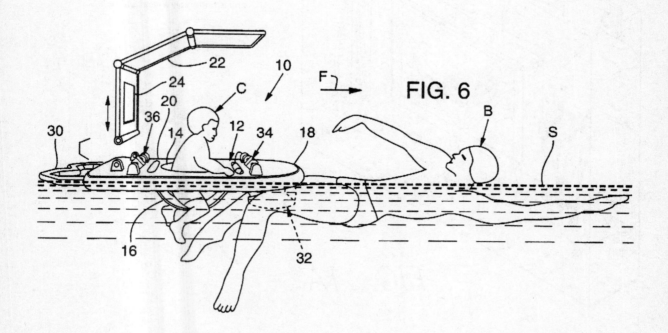

FIG. 6

Patent No. 6,415,739
July 9, 2002
James A. Orendorff; Cory L. Amos

Combination Fish Tank and Picture Frame

Abstract

A tank is combined with a picture frame. The tank has a housing with a recessed portion extending into the interior of the tank. The recessed portion is preferably unitarily formed in a front panel of the tank. The recessed portion is reinforced. The tank preferably has a flat bottom to be used on a table top, and has mounting brackets so it may be hung on a wall. The tank may serve as a fish tank, or other tank, and is preferably a clear plastic.

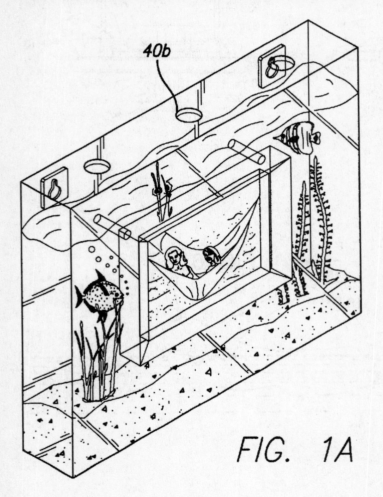

FIG. 1A

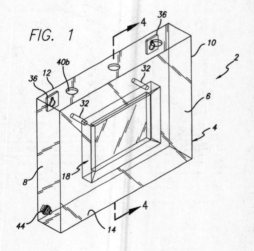

FIG. 1

44

Patent No. 6,024,104

Feb 15, 2000

Jeffery Dolberry

Combination Writing Utensil and Floss Dispenser

Abstract

A combination writing utensil and floss dispenser for writing and dispensing floss. The combination writing utensil and floss dispenser includes an outer housing with separable dispensing and writing portions, opposite dispensing and writing ends, a longitudinal axis extending between the ends, and a marking device disposed in the housing and that is partially extendable through an opening in the writing end. A spool is rotatably or fixedly disposed in the dispensing portion. The spool has floss wrapped around it, a length of which extends through an aperture in the dispensing end of the housing. The dispensing end has a cutting blade for cutting the floss.

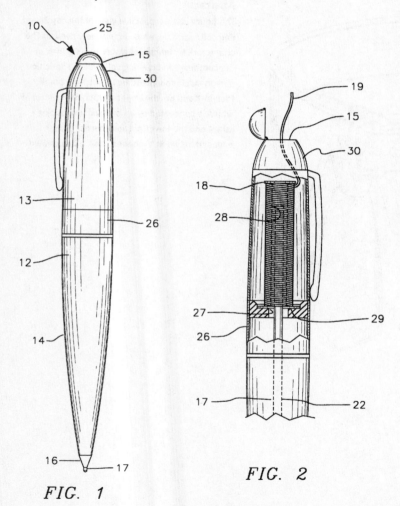

FIG. 1

FIG. 2

Patent No. US D131,052

Jan 6, 1942

Morris Oppenhaim

Combined Dog Collar and Tie

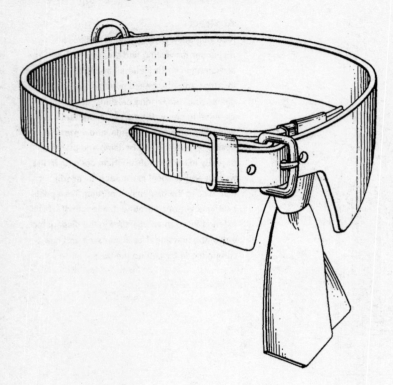

Abstract

The figure is a perspective view of the combined dog collar and tie, showing my new design. The characteristic feature of my design resides in the combination on a dog collar of the specific design disclosed including a relatively small four-in-hand tie, the upper portion of the knot of which is concealed by an overlapping buckle strap, and the lower portion thereof being exposed between V-shaped tabs, all as shown.

Patent No. 6,439,419
Aug 27, 2002
James S. Darabi

Combined Drinking Cup and Horn

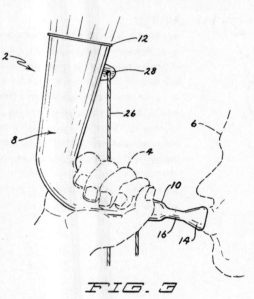

FIG. 3

Abstract

A combined drinking cup and horn comprises a generally L-shaped cup portion having a small diameter first end and a larger diameter second end. The second end of the cup portion is large enough to permit a user to drink from the cup portion through the second end thereof when a lid on the second end of the cup portion has been removed to expose a beverage contained in the cup portion. A mouthpiece is attached by a reduced diameter neck to the first end of the cup portion. An opening in the mouthpiece is normally closed by a removable stopper. When the cup portion is empty of the beverage, the user can then remove the stopper. The user can then blow through the mouthpiece to create a tonal sound similar to that provided by a horn. The combined drinking cup and horn could be sold at sporting events or the like for use at the event.

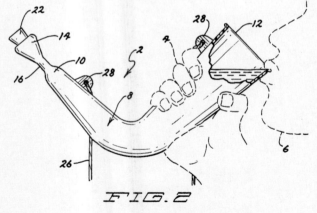

FIG. 2

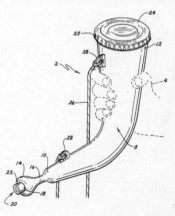

Patent No. 5,740,814
April 21, 1998
Roger Comi

Condom in a Nut Novelty

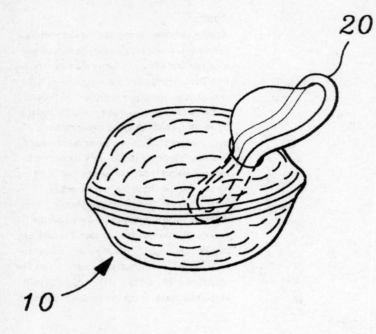

20

10

Abstract

A condom in a nut novelty, wherein a condom is contained within a natural nut. The nut is drilled to create a hole. A blade is inserted through the hole to quarter and then macerate the nut meat. The macerated nut meat is removed through the hole. A condom is inserted into the nut shell by introducing a piece of the condom into the hole, and then repeatedly twisting the condom around the hole and pushing the condom into the hole. The hole is sealed, and then painted to match the nutshell. The nut is cracked by the novelty victim to reveal the condom.

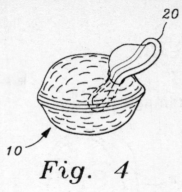

Fig. 1

10 12

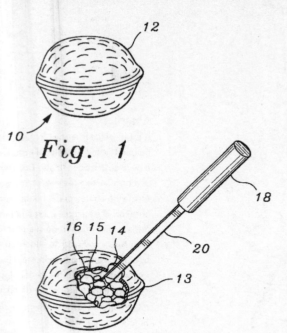

Fig. 2

16 15 14 18 20 13

Fig. 3

14

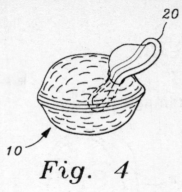

Fig. 4

10 20

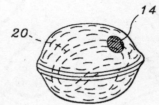

Fig. 5

20 14

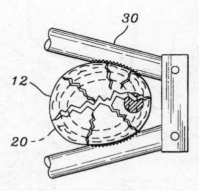

Fig. 6

30 12 20

Patent No. 5,203,708
April 20, 1993
Nancy A. Bird

Creatress Image

Abstract

A multi-piece image of female creation that can be used as toy, archetypal image, educational device or therapeutic aid. The image is made up of a human-like female form together with forms which depict the nonhuman elements of creation of a particular creation myth. The female form may comprise a rigid statue, with a womb-like cavity in its abdomen, with an opening thereto between the legs of the female form. Alternatively, the female form may be a soft figure such as a stuffed doll, with an expandable torso and cavity achieved e.g. by pleating or gathering.

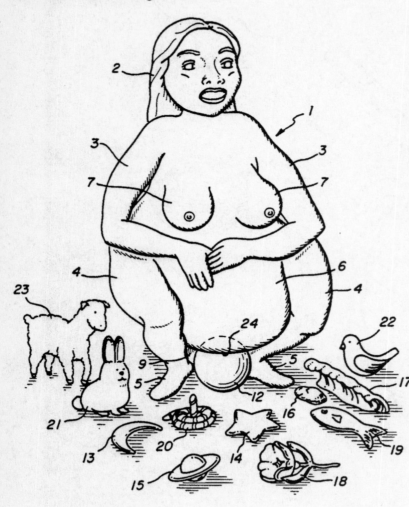

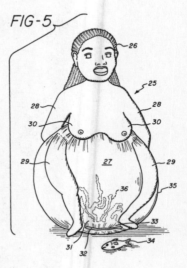

FIG-5

FIG-1

FIG-2

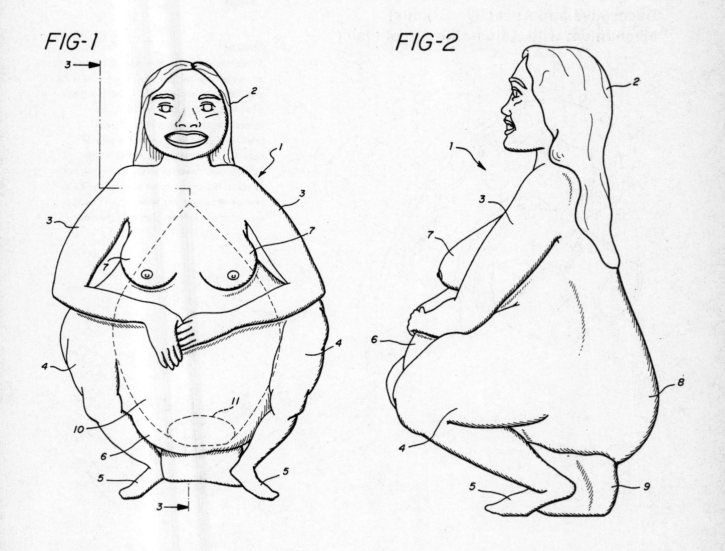

51

Patent No. US 6,694,651

Feb 24, 2004

Shun Tian Shuen

Decorative and Attractive Exhaust Mechanism with Adjustable Angels [sic]

Abstract

A decorative and attractive exhaust mechanism with adjustable angles consists of one rotary multi-direction slip-on assembly that extends along with the size of the exhaust, one decorative plate going with the user's preferences and requirements to provide decoration and attraction and one plastic exhaust cover installed on one end of the multi-direction slip-on assembly so as to provide various decoration and attraction and could be applied to all kinds of automobile exhausts.

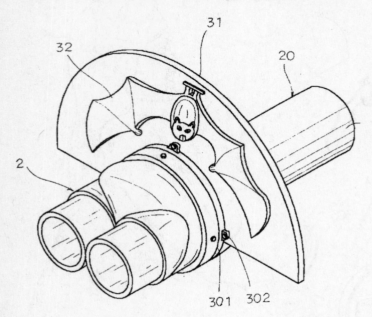

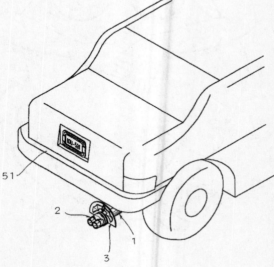

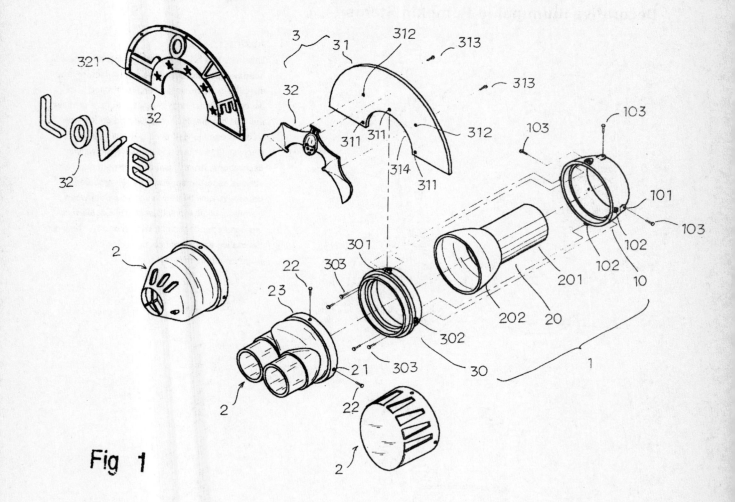

Fig 1

Patent No. US 6,513,945
Feb 4, 2003
John Raymond Wyss; Wade Garrett Latourette

Decorative Illuminated Pumpkin Stems

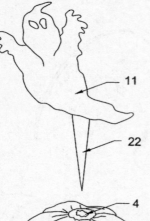

Abstract

This invention is in the area of three-dimensional holiday decorations and their manufacture, and specifically in the area of an illuminated decoration which may replace natural stems for the tops of pumpkins, squash and gourd-type fruits as used in Halloween, Thanksgiving, Harvest, Christmas, and similar holiday-type decorations. An artificial replacement stem and various assorted appendages are provided. A decorative novelty item, such as a Halloween pumpkin, is affixed with said artificial stem or various appendages of a design or other festive decorative appurtenance, replacing or augmenting its natural stem.

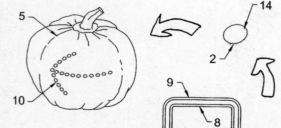

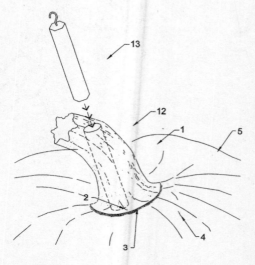

Patent No. 5,950,888

Sept 14, 1999

Patricia Nolan-Brown

Detachable Activity Flap

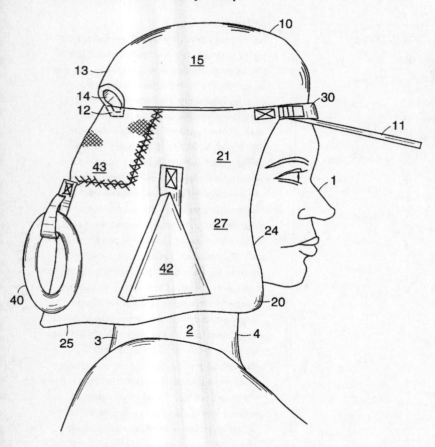

Abstract

A thematic activity neck flap which may be attached to a cap or about the neck of a backpack wearer. The flap is functional and comfortable for a wearer as well as providing a thematic ambiance. The flap has thematic items attached thereto such as pediatric items comprised of shapes, teethers and hanging toys. The flap may also have different textures consistent with a pediatric theme.

Patent No. US 4,088,315
May 9, 1978
Robert Archer Schemmel

Device for Self-Defense Training

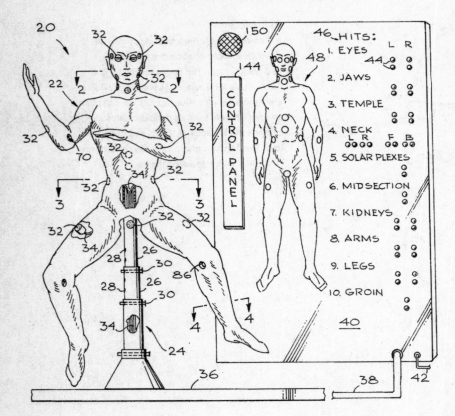

Abstract

An improved device for use in self-defense training, as in karate and the like, includes a lifelike articulated training dummy supported in an upright position on a post and having a plurality of separate pressure receptors disposed at various target locations in the dummy. The receptors are interconnected to a signal such as individual lights in a remote display panel so that hits on the receptors can be separately displayed by the panel. The panel can include a timer, hit sequence counter, hit sequence programmer, printed readout, and hit sequence replayer, as well as a warning signal, visual and/or audible, and other safety training aids. The receptors can be made to distinguish between light and heavy blows. The support post can be rotated at high speed to cause the dummy to simulate an attack when activated by weight detectors in a base around the post. The weight detectors are also disposable in the base in a mode to facilitate stance training. The dummy and post can be provided with shock-absorbing elements to protect them from heavy hits during practice. Preferably the dummy includes a tough, resilient surface layer for further protection of the dummy and trainee (one using the dummy) and for toughening the hands of the trainee. The device provides unique advantages in the art of self-defense training.

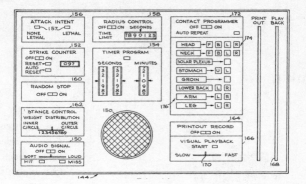

ATTACK INTENT	RADIUS CONTROL	CONTACT PROGRAMMER

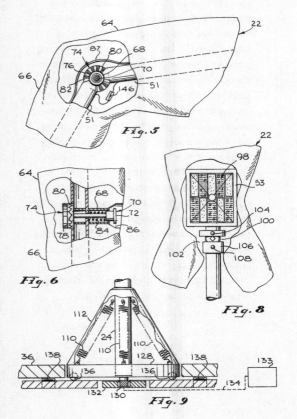

Fig. 5

Fig. 6

Fig. 8

Fig. 9

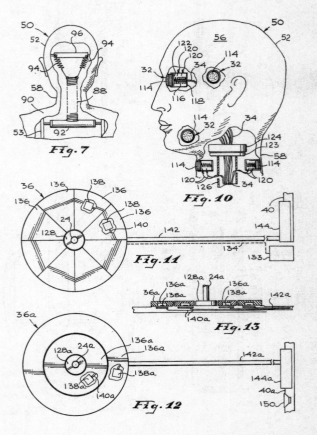

Fig. 7

Fig. 10

Fig. 11

Fig. 13

Fig. 12

Patent No. 3,630,172
Dec 28, 1971
Marcel Neumann; Burton L. Siegal

Diet Reminder Manikin

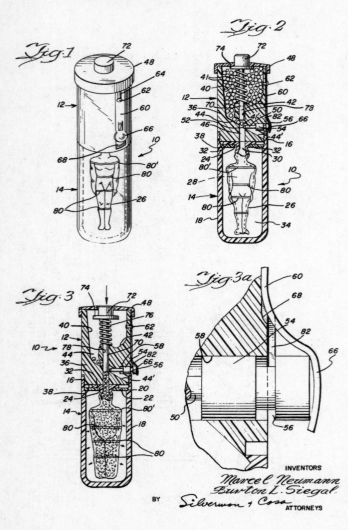

Abstract

A structure comprising a manikin enclosed in a suitable enclosure is capable of manipulation by the user each time that a number of calories of food or grams of carbohydrate has been consumed by the user. The manipulation may take the form of dropping balls into the hollow of the manikin, or releasing a small amount of colored liquid to flow into the manikin or moving a colored member to change the appearance of the manikin. In each case the manipulation is performed during the day in such degrees that normal food intake by the user with orthodox manipulation will not result in any distortion or out-of-the-ordinary appearance of the manikin. An excess of manipulation on the other hand will make a visible change from the normal in the appearance of the manikin. This provides a reminder for the normal food consumption of the dieting user, but also provides a psychological reminder of the results of excess. For example, the manikin will appear bloated, or excessively colored, etc., if the user indulges in excess food and is religious in manipulating the manikin for each increment of intake.

INVENTORS

Marcel Neumann
Burton L. Siegal

BY Silverman & Cass ATTORNEYS

Patent No. 5,807,301

Sept 15, 1998

Igal Nadam

Disposable Device for Safe Cleaning of the Ear

Abstract

A disposable ear cleaning device includes a thimble having a rigid extension protruding from a top portion thereof with a layer of cotton or other soft material wrapped around the protrusion.

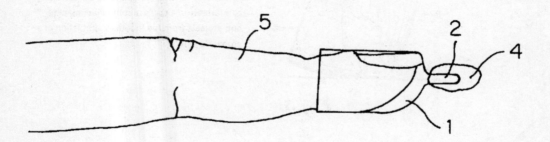

Patent No. US 2,610,851

Sept 16, 1952

Win Jones

Dog's Toy

Fig.1

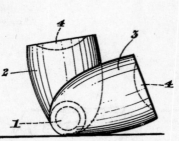

Fig.3

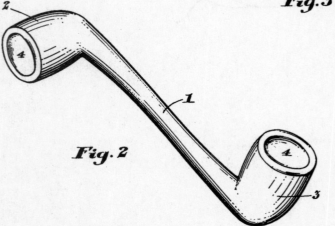

Fig.2

Abstract

The present invention relates to toys for dogs and more particularly to dog's toys which are molded of rubber or other resilient material. The primary objects of the invention are to provide a dog's toy which, when held in the mouth of a dog or puppy, will produce a comical effect as though the animal were smoking a pipe; to provide such a toy which is molded of rubber or the like, and which is so formed that either end thereof will fit comfortably in a dog's mouth while the other end projects from his mouth in simulation of a smoker's pipe.

Patent No. 5,456,625
Oct 10. 1995
Linda M. S. Drummond

Dolls Formed in the Likeness of the Lord Jesus with a Movable Head and Extremities

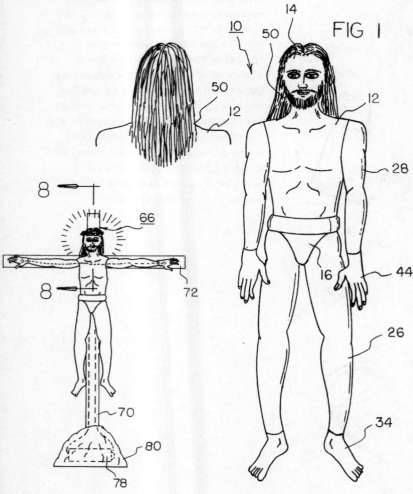

FIG 1

Abstract

A doll formed in the likeness of the Lord Jesus with a movable head and extremities comprised of a torso section including a loin cloth molded into its lower portion and a pair of movable leg sections. The lower ends of the leg sections have a pair of feet sections secured thereto. A pair of arm sections are secured to the torso section. A pair of hand sections are secured to the lower ends of the arm sections. A head section is secured to the upper end of the torso section. The doll is provided with electrically conductive nails which when inserted through apertures in the hands of the doll, mount the doll to a provided cross and close an electrical circuit which illuminates the cross.

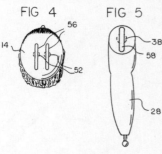

FIG 4

FIG 5

FIG 6

Patent No. 6,048,209
April 11, 2000
William V. Bailey

Doll Simulating Adaptive Infant Behavior

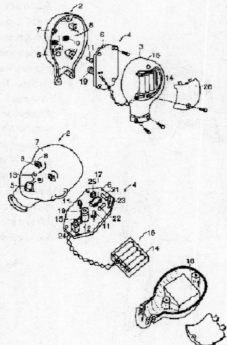

Abstract

A doll simulating infant behavior includes sensors that detect care given to the doll, such as feeding, rocking, and neglect or abuse, and provide inputs to a microcontroller which operates on a behavioral state machine to produce infant behaviors that are expected of a human infant in response to similar care. The sensors include a feed-switch, motion sensors, and an impact sensor. LED eyes, a speaker, and a bi-colored blinking LED provide the doll's heartbeat and overall health. The microcontroller causes the doll to undergo a plurality of behavioral cycles such as sleeping, hunger, feeding, crying, wailing, colic and burping.

Patent No. 4,136,483
Jan 30, 1979
Judy Shackelford; Rouben T. Terzian

Doll with Changeable Facial Features

Abstract

A figure toy, such as a doll or the like, having a torso and a head. The head has a brow and scalp portion secured to the torso by a vertical post and against rotation relative to the torso. Between the brow and scalp portion and the torso, the head includes a face section or portion mounted for rotation on a vertical axis relative to the scalp portion and torso when the doll head is considered in upright position, e.g. on the axis of said post. The face portion includes back-to-back different facial features so that on rotation the front facial features can be changed. In order to conceal the facial features rotated to the rear of the head, the scalp portion is provided with simulated hair extending therefrom downwardly at the rear of the head to a position below said facial feautures.

Patent No. 4,249,338

Feb 10, 1981

Howard Wexler

Doll with Sound Generator and Plural Switch Means

Abstract

A doll comprising a crying sound generator within the doll body, first switch means for actuating the crying sound generator, at least two additional switch means, an automatic selection means for determining which of the additional switch means is connected to stop the crying. The additional switch means may be operated by manipulating the doll, e.g., by giving it a bottle, by changing its diaper, or by picking it up and patting its back. A short sequence of sighing or cooing sounds may be used at the end of the crying sounds.

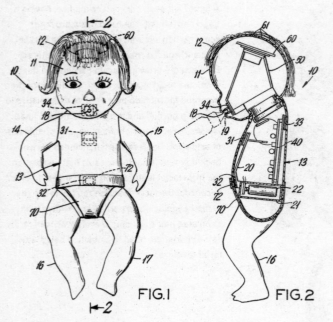

FIG.1

FIG.2

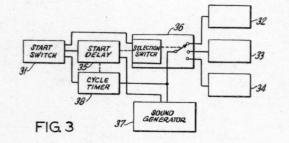

FIG. 3

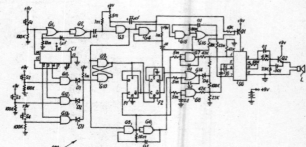

Patent No. 2,636,176

April 28, 1953

Howard C. Ross

Double Coat

Abstract

The invention relates to garments of the type generally referred to as top coats or rain coats, and in particular this invention relates to an extensible coat wherein, in an emergency, the coat may be extended to include two persons. Various attempts have been made to provide protecting garments for use in an emergency such as in an athletic stadium or on a golf course but it has been found difficult to provide a combination garment for this use that could also be used for walking and that is also adapted for universal use. Another object of the invention is to provide an extensible garment which in the collapsed or folded position has the appearance of a conventional coat.

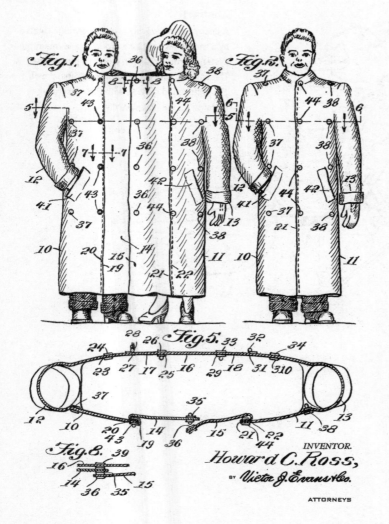

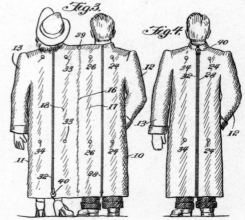

INVENTOR.

Howard C. Ross,

BY Victor J. Evans & Co.

ATTORNEYS

Patent No. US 6,493,980
Dec 17, 2002
Johnny J. Richardson; Barton D. Richardson

Duck Decoy with Quick Release, Foldable Wings

Abstract

A decoy has outwardly projecting wings rotated by a motor secured by a modularized mounting system shrouded within a lifelike plastic shell. The wings rotate about their longitudinal axis. The wings stop moving with their bright white bottoms aimed downwardly.

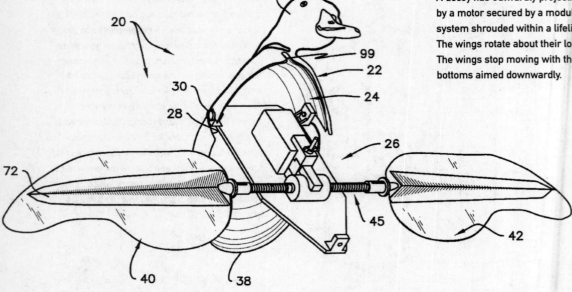

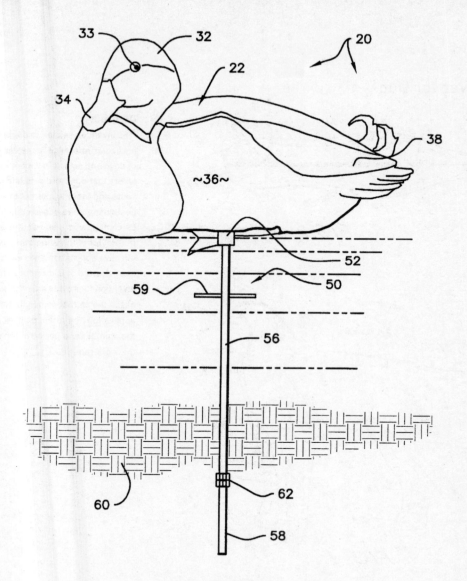

Fig. 2

Patent No. US 3,150,641
Sept 29, 1964
Seroun Kesh

Dust Cover for Dog

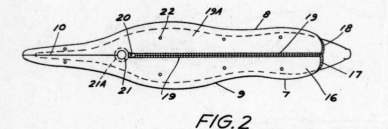

FIG. 2

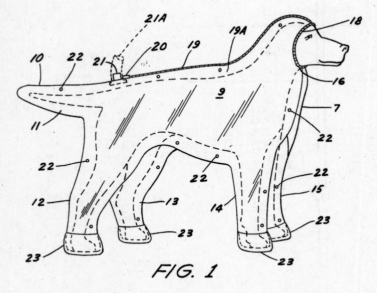

FIG. 1

Abstract

This invention relates to domestic pets and has particular reference to the care and well-being of dogs and cats. It is well-known to animal lovers that dogs and cats harbor fleas and other pests and that the attempt to eradicate such pests oftentimes poses quite a problem. Efficient powders and sprays are now on the market for eliminating these pests, but their effective application and retention leaves much to be desired. It is therefore an object of this invention to provide a device, which will greatly assist the animal owner, in the effective application and retention of such pesticides on the animal, for a period of time sufficient to do the work required.

Patent No. D400,467

Nov 3, 1998

Chris Campbell

Easter Tomb Pendant

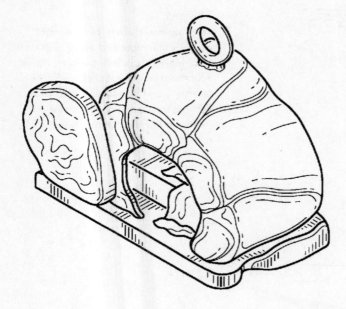

Abstract

The ornamental design for an Easter tomb
pendant, as shown and described. This is a
perspective view of an Easter tomb pendant
embodying the present invention with a movable
element in the first position.

Patent No. 2,501,902

Mar 28, 1950

Judy Bellamy Howell

Educational Toy

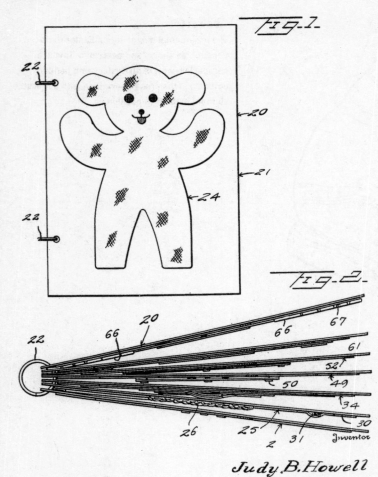

Abstract

This invention relates to toys and more particularly to educational toys in book form. It is an object of this invention to provide an educational toy of the kind described, for teaching a child the various processes of securing together articles of clothing in the various manners conventionally applied.

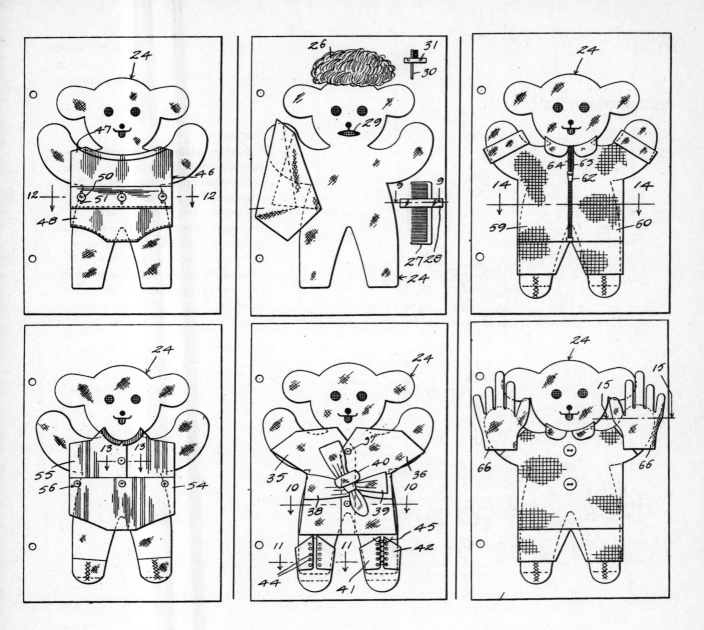

71

Patent No. 1,496,406

June 3, 1924

Marguerite Bertsch

Expression Doll

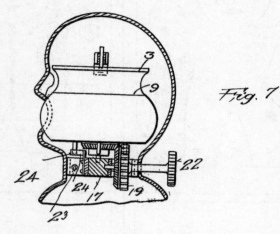

Fig. 7

Abstract

The invention relates to expression dolls. The object of the invention is to provide, in a doll's head, novel means to produce a multiplicity of expressions of the doll's face. A further object of the invention is to provide certain eye expressions which correspond to certain mouth expressions, denoting joy, anger, fear, sorrow, etc., and means whereby the combined corresponding eye and mouth expressions are successively disposed through apertures in the doll's face.

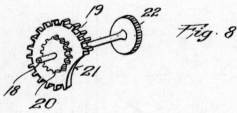

Fig. 8

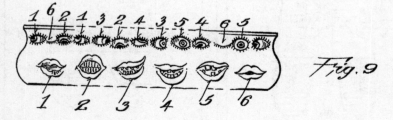

Fig. 9

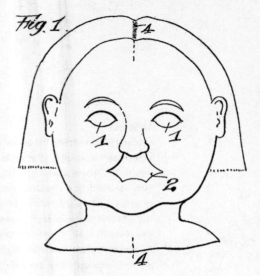

Fig. 1.

4

1 1

2

4

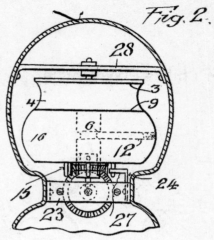

Fig. 2.

28

3
9

4

16

6

12

15

24

23

27

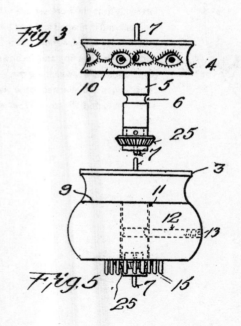

Fig. 3.

7

4

10

5

6

25

7

3

9

11

12

13

Fig. 5.

7

15

25

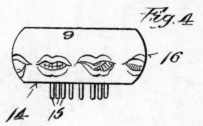

Fig. 4.

9

16

14

15

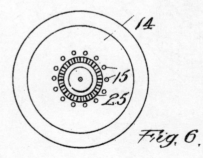

14

15

25

Fig. 6.

73

Patent No. 5,022,666

June 11, 1991

Gregory L. Simon

Facade for Child's Play Vehicle

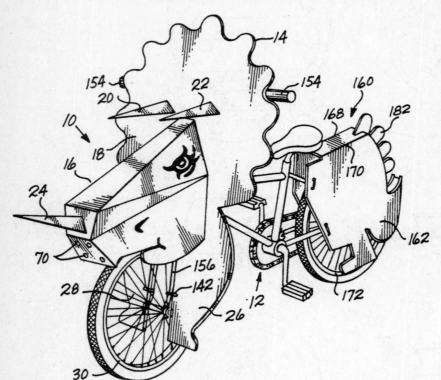

Abstract

A facade for simulating the appearance of another object which is attached to a child's play vehicle. It includes a first substantially planar member which has a front surface and is attached to a forward portion of the play vehicle. The first member is shaped to have a perimeter which substantially defines an anterior profile of the object. A second substantially planar member is folded into a three-dimensional planar member. The second member projects forwardly from the front surface of the first member and is shaped to define a forward profile of the object. It may also include a rear portion which has first and second substantially planar side panels which are connected along upper edges and attached to the play vehicle in a position to partially cover a rear wheel of the vehicle.

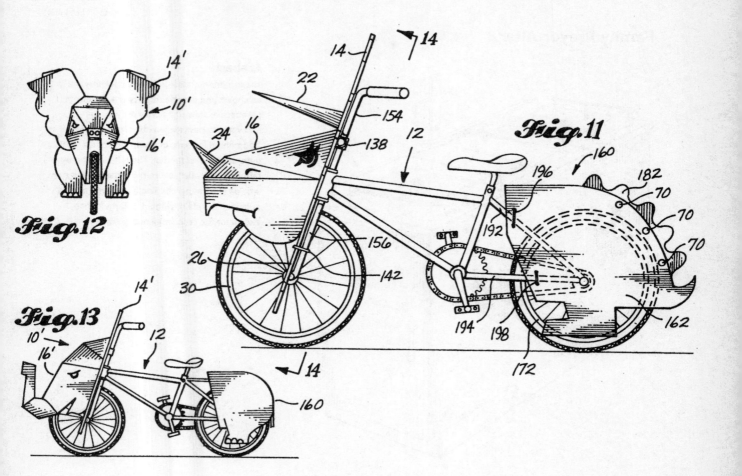

Fig.12

Fig.13

Fig.11

Patent No. D245,381
Aug 16, 1977
Ebert Lee Lilley

Family Prayer Altar

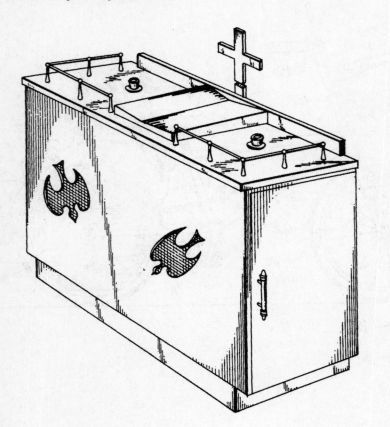

Abstract

The ornamental design for a family prayer altar, as shown and described. This is a right front perspective view of a family prayer altar showing my new design, the cross shown fragmented for ease of illustration and a front perspective view thereof. The left side of my family prayer altar is a mirror image of the right side as shown therein. The back of the altar is substantially plain, without ornamentation, except for the cross attached thereto.

Patent No. US 6,571,399
June 3, 2003
Reuben O'Neil Wagener

Female Urinal Fixture

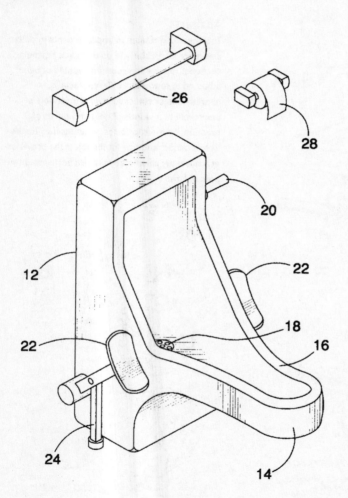

Abstract

A porcelain or stainless steel urinal bowl that mounts to the wall and has a narrow front extension out from the bowl, which will allow a female user to have convenient access above the bowl. The narrow front part of the bowl will be no more than six inches wide so the user can easily straddle the bowl. The bowl has a fresh water supply, which supplies water through a plurality of holes underneath the upper rim opening of the bowl, and a flush mechanism, which has a series of drain holes that empty urine and water into the connected sewer drain when flushed by means of a hand or foot lever mounted to the bowl.

Patent No. 1,762,374
June 10, 1930
William O. Yancey

Figure Doll

Abstract

This invention relates to improvements in dolls and more particularly to dolls' heads having a sectional rotative face portion capable of being adjusted to reproduce different facial expressions or contortions. The invention is applicable to all kinds of dolls or figures of persons, animals, or birds, whether they be life-like or comic, and has for its object the provision of a novel toy calculated to afford amusement to both old and young.

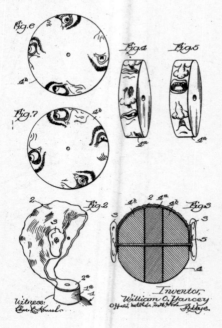

Patent No. 5,848,928

Dec 15, 1998

Ken E. Wong

Finger Puppet Eating Utensil

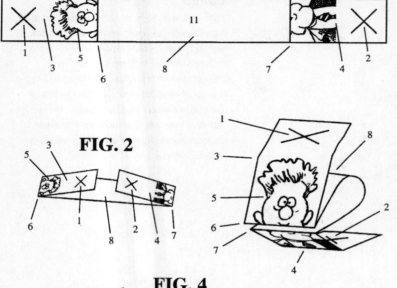

FIG. 2

FIG. 4

Abstract

A simple and inexpensive combination grasping tool and puppet is disclosed in the form of a sheet with distal ends. The tool comprises a single sheet of pliable material with finger-lock "x" slits that provide a custom locked fit around a user's fingers that have been pushed through the slits, thus freeing the user from the requirement of holding onto the tool. As the tool is worn and manipulated between the thumb and index finger, strategic placement of artwork or other indicia exemplifying a face upon the surface of the sheet then causes the device to appear to be an articulate puppet which can open and close its mouth.

Patent No. 6,128,863

Oct 10, 2000

George D. Millay

Fish and Marine Mammal Observatory Featuring a Carousel that Moves Within a Sealed Aquatic Environment

Abstract

A fish and marine mammal aquatic observatory. The observatory comprises an outer cylindrical wall and inner cylindrical wall that define an enclosed annular volume. The enclosed annular volume is partially filled with a body of seawater that includes fish and marine mammal animals and other aquatic plants. The observatory further includes a carousel or platform supported on a tower located along a longitudinal axis of the observatory.

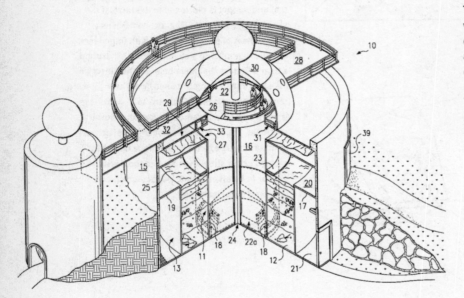

Patent No. US 6,443,787
Sept 3, 2002
Robert C. Woolley

Flying Ski

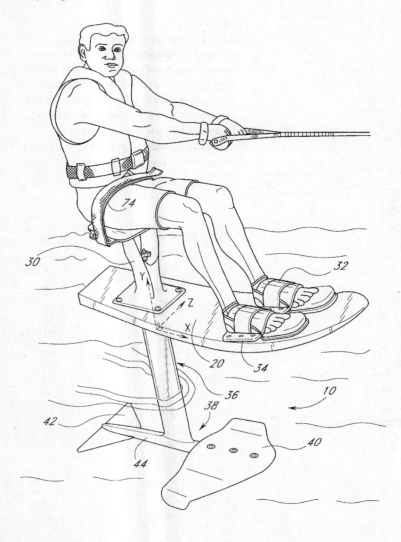

Abstract

The improved flying ski is designed to be towed behind a conventional powered watercraft utilizing a standard ski tow rope or similar device having a handle that can be held by a human rider. In use, the rider is seated on the seat of the flying ski and towed by the watercraft. The improved flying ski comprises an elongate board and a seat that extends generally perpendicular to and upward from the board to support the seated rider's buttocks. The improved flying ski accommodates a variety of rider skill levels by incorporating a mechanism and system that allows the rider to selectively adjust performance characteristics of the ski. In particular, the rider can control stability, lift, and maneuverability ski characteristics to accommodate the rider's particular skill level and the particular challenge that the rider seeks. In addition, the flying ski includes a detachable back support that allows handicapped riders to enjoy the thrills of using the ski.

Patent No. US 6,387,064
May 14, 2002
Brent Gunnon

Foot Pump Powered Neck Massaging Device

Abstract

A foot powered neck massaging device including a first bladder and a second bladder securable to a lower surface of a shoe worn by a user. The first bladder and the second bladder each have an air outlet. A massaging hand is in communication with the first bladder and the second bladder. The massaging hand is comprised of pliable fingers. The pliable fingers are conformable around the user's neck.

Fig. 1

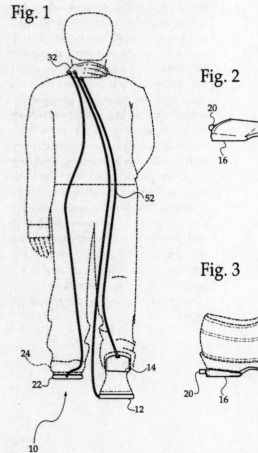

Fig. 2

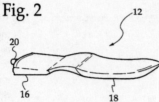

Fig. 3

Fig. 4

Fig. 5

Fig. 6

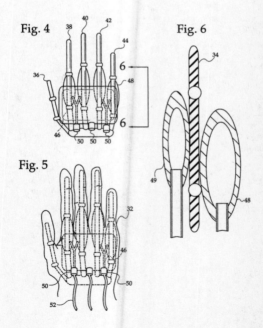

Fig. 7

Fig. 8

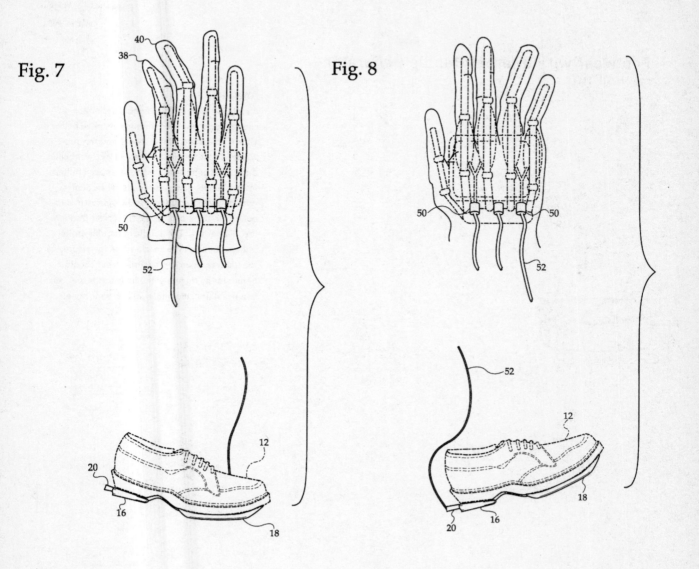

Patent No. 6,035,553
Mar 14, 2000
Lynn Mercier

Footwear with Internal Bubble Generator

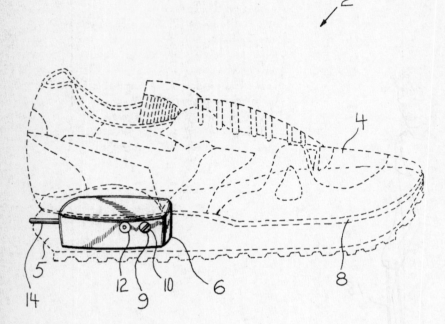

Abstract

A footwear with integral bubble generator comprising a bubble generator installed below a shoe insole. When the footwear wearer puts weight on the insole, the insole presses against the bubble generator, and bubbles are emitted from a bubble generator reservoir through a nozzle. Bubble solution may be poured into (or out of) the bubble generator reservoir through the fill aperture. The bubble generator emits bubbles through the nozzle when the wearer of the footwear exerts pressure on the bubble generator by means of the footwear insole, as when walking, jogging, running, dancing, etc.

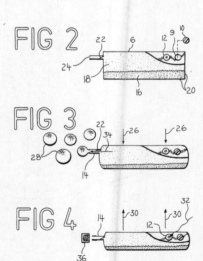

Patent No. 3,933,353
Jan 20, 1976
Charles M. Marsh

Game Device for Two or More Persons Which is Used with the Feet

Abstract

A game device which includes a pair of rigid strip members such as planks, each carrying at least two upwardly projecting loop parts. With the strip members placed in parallel fashion upon the ground two or more players stand upon the strip members, grasping the bight of the loop parts with each hand. The players pull upwardly on the loop parts, causing the strip members to be held against the feet, and walk or shuffle the strip members across the ground.

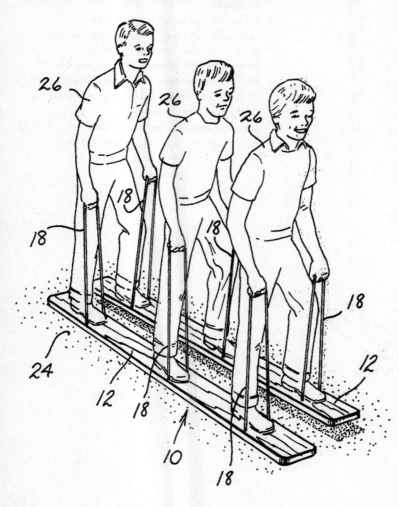

Patent No. US D207,493

July 11, 1966

Robert J. Mader

Golf Putter

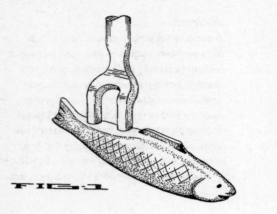

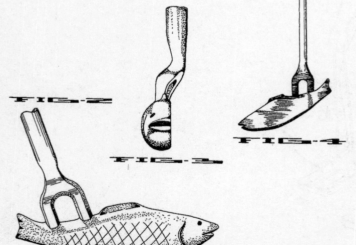

Abstract

The ornamental design for a golf putter, as shown and described. This is a perspective view of the putter taken from one side thereof, showing my new design. Also a side elevation view of the putter head, an end elevation view of the putter head, and a side elevation view of the putter taken from the side thereof opposite to that of the view. The shaft has been shown fragmentarily for convenience of illustration.

FIG-1

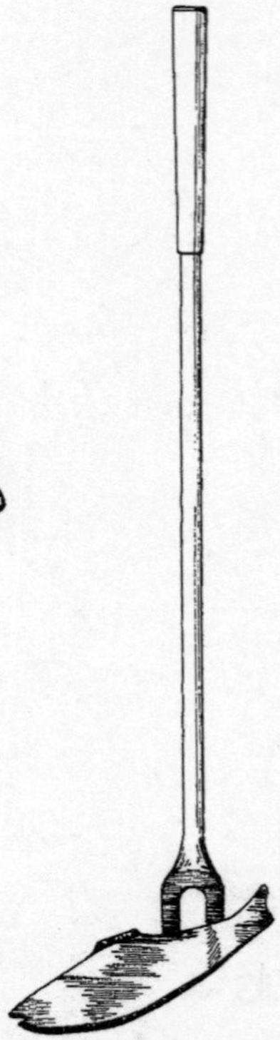

FIG-2

FIG-3

FIG-4

Patent No. 4,563,981

Jan 14, 1986

Roy L. Kramer

Group Tether Apparatus

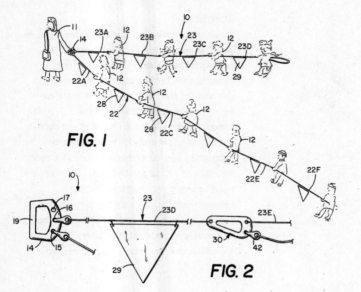

FIG. I

FIG. 2

Abstract

A group tether or control apparatus for imparting directional control to a group of individuals, such as children, on an outing or field trip, as when crossing a street or touring a building. The group operator has control of a main handle. The group leader leads the group by walking ahead with the main handle, and the group members follow by maintaining a grasp on the secondary handles.

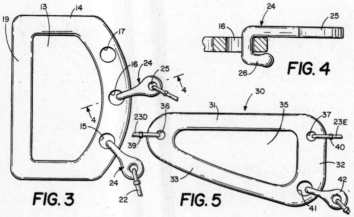

FIG. 3

FIG. 4

FIG. 5

Patent No. US 6,435,931
Aug 20, 2002
Robin Marie Yaeger; Roy Robert Ryer

Group Walking Toy

Abstract

A mobile toy having a plurality of compartments for containing persons. At least some of the components are sized to contain walking children. Each compartment is defined by a sidewall structure having a side opening for ingress and egress and each of the compartments is open at a top and at a bottom. The toy is further provided with means for suggesting an identity that stimulates a child's interest and includes structure for utilizing walking energy of persons within the compartments to cause the toy to move. The toy permits children to be organized for walking from one location to another while simultaneously enjoying a play experience.

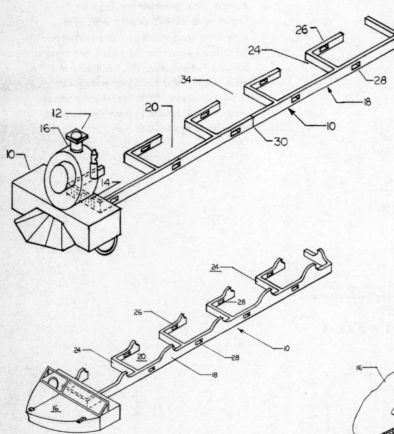

FIG. 2

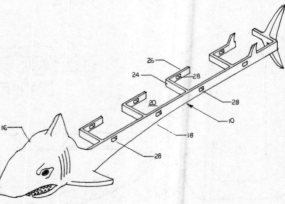

FIG. 3

Patent No. US 6,585,387
July 1, 2003
Chun Teng Lee

Hanging Bubble Lamp

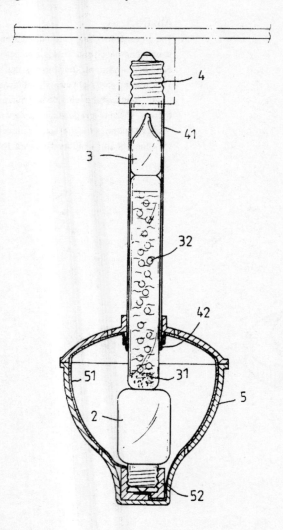

Abstract

The present invention relates to a hanging bubble lamp, which includes a conductive receiver with a sleeve to receive a water tube therein and two conductive wires extend downward. The ends of the tube and the sleeve are engaged within a holder, which has two conductive slices to electrically contact with the conductive wires and a heating bulb in a connector in the holder. By use of the conductive, the water tube of the bubble lamp will be kept in a hanging position for providing a special decorative effect.

Patent No. D401,202
Nov 17, 1998
Thomas T. Washington Sr.

Hood Ornament

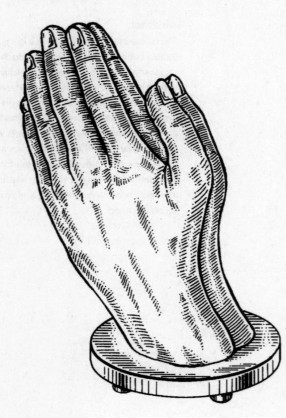

Abstract

The ornamental design for a hood ornament, as shown and described. This is a perspective view of a hood ornament, showing my new design. This is a reduced right side view thereof, it being understood that the left side is a mirror image thereof. There is a reduced front view thereof, a reduced bottom view thereof, a reduced rear view thereof and a reduced top view thereof.

Patent No. 5,769,724

June 23, 1998

Theodore F. Wiegel

Human Free-flight Catapult

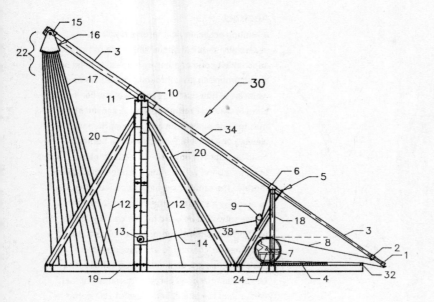

Abstract

An amusement ride for catapulting a human rider, enclosed within a capsule, into the air in much the same fashion as ancient armies would hurl large boulders over the walls of castles. The ride essentially consists of a siege type catapult and a releasably connected launch pod. After the launch, the rider is catapulted at a physiologically safe rate of acceleration along a predictable free-flight arc. When an acceptable elevation is reached, the rider is separated from the capsule and gently brought to earth using an automatically deploying parachute or similar device. An alternate embodiment envisions the use of a similar device to project a conveyance vehicle along a horizontal track similar to a stone skipping the water.

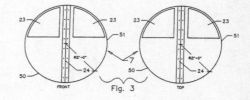

Fig. 3

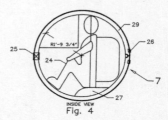

Fig. 4

Patent No. US 6,668,749
Dec 30, 2003
William H. Fargason

Hunting Accessory

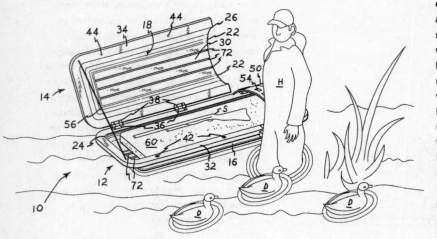

Abstract

A hunting accessory comprises a pair of essentially identical shells which are removably attachable to one another by hinges. At least one shell comprises an unbroken, continuous and waterproof hull useful as the lower portion of a blind in water or marshy areas. A second shell may include a separate openable portion with viewing openings disposed over the hunter's head and upper body when the assembly is used as a blind with a supine hunter concealed therein. The smaller portion is opened when the hunter has a shot at game in the vicinity. The device may also be used for the carriage of hunting equipment, and/or is useful as a closed container for containing and hauling loose materials (leaves, etc.) in an open vehicle. The two shells may be separated from one another, with one of the shells being useful as a game drag or the like.

Patent No. 56,413

July 17, 1866

Charles Hess

Improved Combined Piano, Couch and Bureau

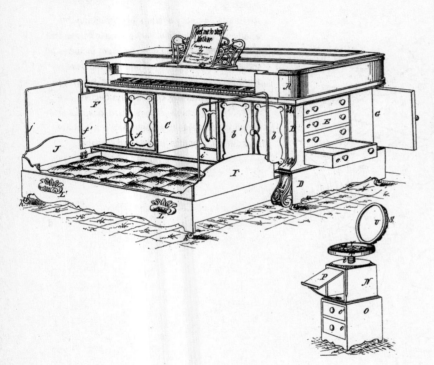

Abstract

My invention relates to the addition of a couch, bureau, &c., to an ordinary piano, these articles being made to fill up the unoccupied space underneath the piano, and being arranged in such a manner as not to detract from the appearance of the instrument or to interfere with its musical qualities. The convertible piano has been designed principally for the benefit of hotels, boarding schools, &c., containing apartments which are used for parlors, &c., in day-time and yet required for sleeping rooms at night.

Patent No. US 6,634,133
Oct 21, 2003
Patricia A. Levandowski

Inflatable Decoy System

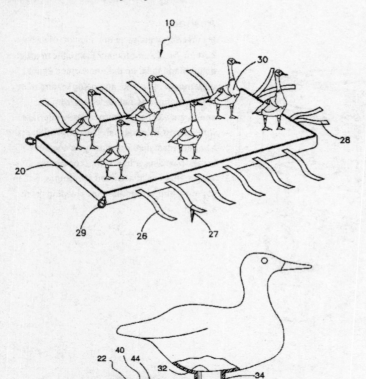

Abstract

An inflatable decoy system for combining the ability of attracting fowl with the ease of use and storage. The inflatable decoy system includes an inflatable base member. The inflatable base member has a plurality of openings, which are coupled to a plurality of inflatable decoys. The plurality of inflatable decoys, each includes an opening couplable to an associated one of the plurality of openings of the base member.

Patent No. US 5,682,701
Nov 4, 1997
Terry G. Gammon

Inflatable Hunting Decoy

Abstract

An inflatable hunting decoy includes a plastic inflatable body form having a plurality of flaps secured on opposing sides thereof. A plurality of grommets are individually disposed within the plurality of flaps of the plastic inflatable body form. A plurality of straps extend through the plurality of grommets disposed within the plurality of straps on opposing sides of the inflatable body form for securement thereof to a tree.

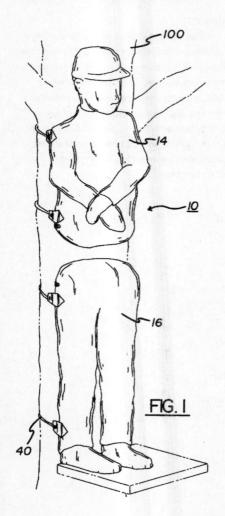

FIG. 1

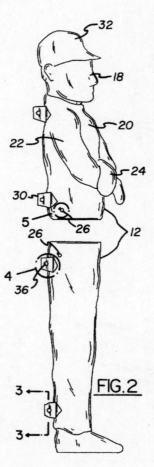

FIG. 2

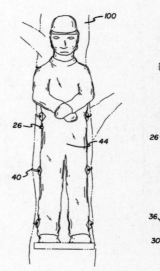

Patent No. US 6,280,344
Aug 28, 2001
Jerry Robb

Luminous Bowling Ball

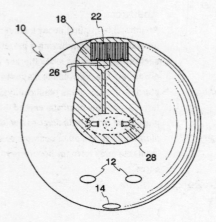

Abstract

An impact illuminated bowling ball including a light transmitting core, a pair of LEDs embedded in the core, a piezoelectric transducer embedded in the core and electrically connected to the LEDs. A shock amplifying mechanism in the form of a steel ball is located in operative engagement with the piezoelectric transducer. A rechargeable electric battery in the core is electrically connected to a solar electric collector for recharging.

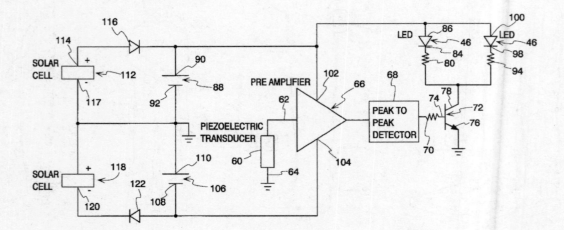

Patent No. US 6,464,654
Oct 15, 2002
Julia S. Montgomery; Sharleen M. Warner

Massaging Shoe Device

Abstract

A massaging shoe device for providing vibrating massage to the bottom portion of a wearer's foot. The massaging shoe device includes a sole having a cavity positioned therein. A vibrator is positioned in the cavity. A power supply for powering the vibrator is electrically coupled to the vibrator. An actuator for selectively turning the vibrator on and off is electrically coupled to the vibrator. A perimeter wall is integrally coupled to a perimeter edge of the sole and extends upwardly therefrom for substantially covering a foot of a user.

Patent No. US 6,447,359

Sept 10, 2002

Carlos D. B. Crump

Memorial Novelty Doll Device Having Integral Sound Producing Means and Kit and Method for Customizing the Same

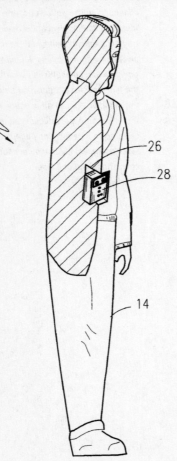

Abstract

A doll is provided that is custom designed to resemble a specific person, either living or dead. These characteristics of sex, hair color and style, eye color and the like are selected from a group of various characteristics. A tape-recording mechanism provides for the recording of the represented person's voice, if still living, or for the recording of voices by friends or relatives, if deceased.

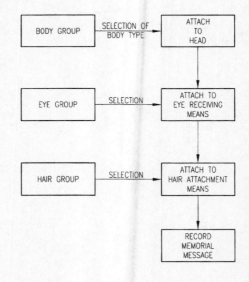

Figure 5

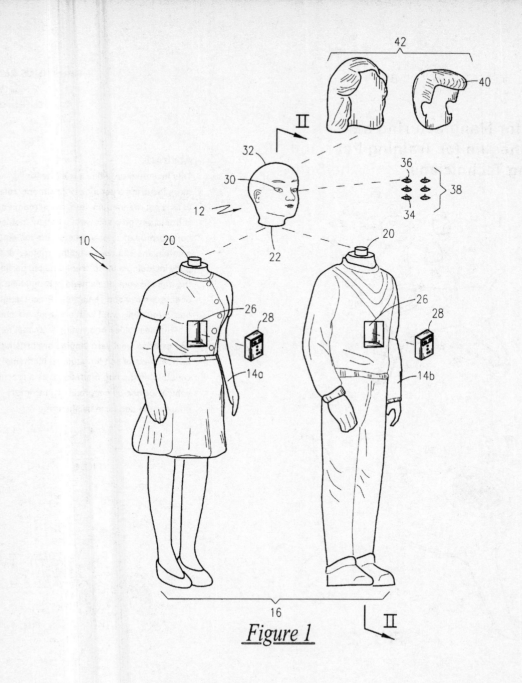

Figure 1

Patent No. US 6,629,872
Oct 7, 2003
Jeremy Chi Kong Cheung

Method for Manufacturing a Pet Mannequin for Training Pet Trimming Technicians

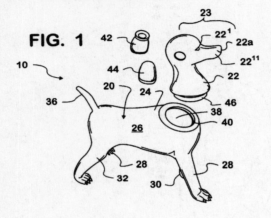

FIG. 1

Abstract

This invention provides a method for manufacturing a pet model, commonly referred to as a pet mannequin, here a dog mannequin, is implanted on a soft walled hollow molded dog body formed of a polyvinyl chloride molding solution, the soft walled hollow molded dog body including a torso, trunk and leg portions, the molded body and a head portion and a pair of ears are molded separately. Precut lengths of hair-simulating wool yarn are implanted into the wall of the molded body using a crochet needle. The exterior wool yarn lengths are brushed to form a mass of soft longitudinal filaments constituting the dog mannequin as a practice vehicle for use by pet grooming trimmers shaping the dog form by shearing.

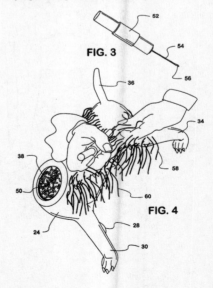

FIG. 3

FIG. 4

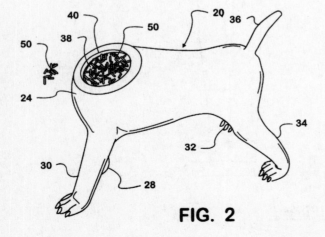

FIG. 2

FIG. 5

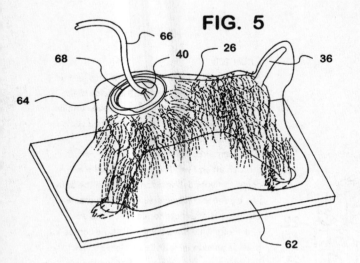

FIG. 7

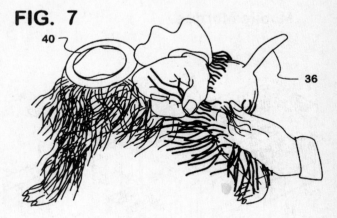

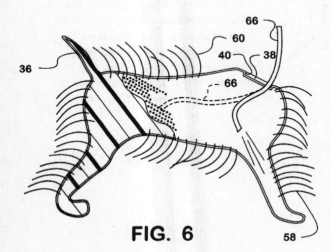

FIG. 6

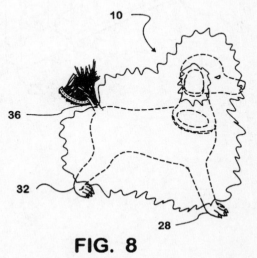

FIG. 8

101

Patent No. US 6,299,229
Oct 9, 2001
Moises Becenas Nieto

Mobile Morgue

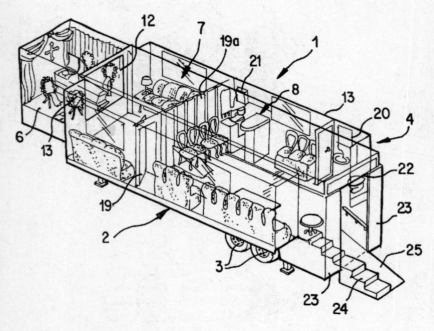

Abstract

A mobile mortuary in which a semi-trailer has extendable large side walls. The inside of the semi-trailer which constitutes a mobile unit is divided into three compartments, one a cooled end cabin with a side access door for emplacement of the casket, which is visible through an end window, and the other two compartments define a zone in which the body has extendable facing side walls. The mobile mortuary is of simple construction and also enormously functional when the time comes to provide an extensive and complete service to the deceased and his/her family and friends. Furthermore, its functionality makes it possible to move the aforesaid mortuary quickly to the home or temporary resting place of the deceased. The nonexistence of mobile mortuaries creates a major obstacle for inhabitants of small communities in providing the necessary services to a deceased person.

Patent No. US 6,675,408
Jan 13, 2004
Cecile L. Mason

Modular Airplane-Shaped Bedroom Furniture

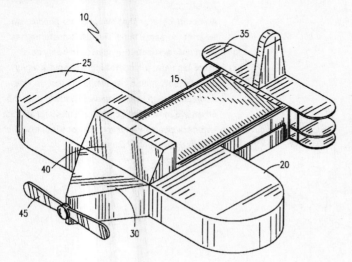

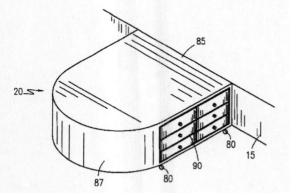

Abstract

Modular airplane shaped bedroom furniture provides a line of bedroom furniture for children that is in the general shape of an airplane. The body of the airplane incorporates a standard twin size mattress. The nose of the modular airplane shaped bedroom furniture incorporates a clothes hamper and wastebasket, while the top of the nose section provides a desk. The single center-mounted propeller is fixed in place so that it cannot turn and provides hanging hooks for clothes, book bags and the like. The wing section on one side of the modular airplane shaped bedroom furniture is a toy box, while the wing on the other side provides six drawers for storage. Under-bed storage provides room for more drawers or an additional roll away bed for use by another child. The tail section can be used as a bookshelf or for storage of shoes. Finally, the cockpit area can be used for storage of a radio, CD player, small television or the like. All components of the modular airplane shaped bedroom furniture are modular allowing for easy moving and assembly. The use of the modular airplane shaped bedroom furniture provides an alternative to conventional children's furniture that is not only unique and eye-catching but fun for children too.

Patent No. 5,970,518

Oct 26, 1999

Aurellius M. Jordan

Multiple Person Outfit

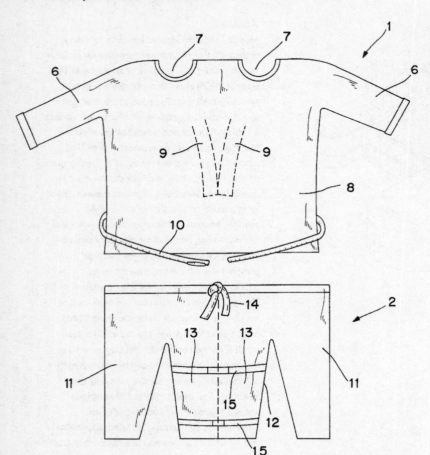

Abstract

An outfit/costume that two or more people will wear at the same time. The outfit/costume has what appears from the outside to be three or more legs and the center leg will hold a leg of each person. It also has two arm sleeves, one for each person, inner sleeves that will hold the other two inner arms of the persons and at least two neck apertures for the persons wearing the outfit/costume.

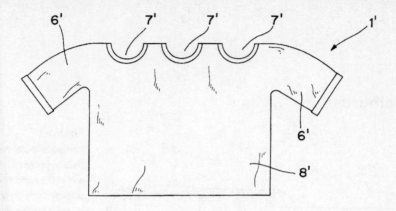

FIG. 3

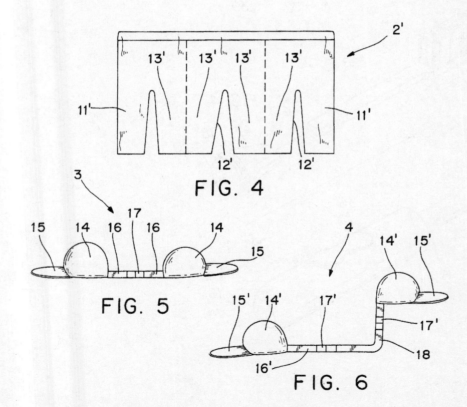

FIG. 4

FIG. 5

FIG. 6

Patent No. US 6,202,245

Mar 20, 2001

Ramin Khodadadi

Musical Toothbrush

Abstract

A musical toothbrush including a handle. A coupling member couples with an open upper end of the handle. A digital sound generator is disposed interiorly of the handle. A speaker is disposed within the handle. An activation button couples with the digital sound generator.

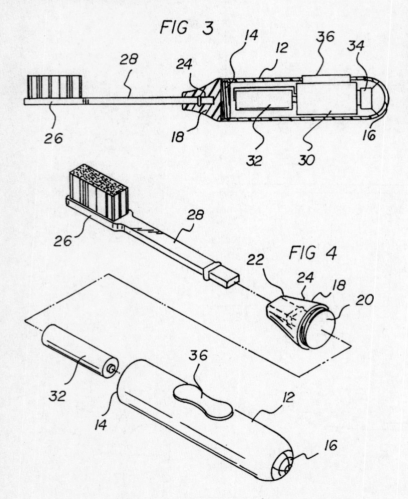

FIG 3

FIG 4

Patent No. US 6,557,728
May 6, 2003
Jeffrey Blake Anderson; Michael Barnett

Musical Toothpaste Tube Closure

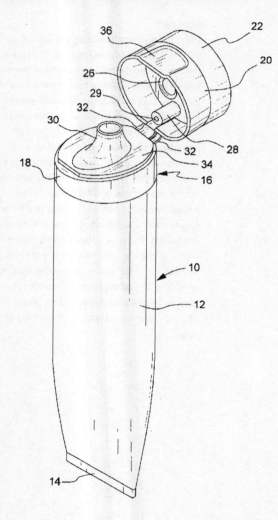

Abstract

A dentifrice closure on a dentifrice tube can emit
a signal upon the opening of the closure to
dispense some of the dentifrice. The closure is
comprised of a base portion to attach the
closure to the dentifrice tube and a lid portion
that closes the dispensing opening of the
dentifrice tube and which contains switch timer
and signal generator. The signal generator can
be a light, but preferably is a sound generator,
and most preferably a music generator. Upon
the lid being opened, a switch activates the
timer which in turn activates the signal
generator. Regardless of the position of the
switch after a lid opening (the lid can be quickly
closed) the signal will be emitted for a set period
of time.

Patent No. US 6,701,924
Mar 9, 2004
Richard D. Land Jr.; Dawn Land

Nasal Filter

Abstract

The nasal filter is a breathing aid that would filter pollutants and contaminants from the air, thus improving the quality of the air inhaled by individuals who suffer respiratory ailments and allergies to air borne pollutants. Designed to fit just inside the nostrils, the nasal filter would consist of a spring clip having flanged two-pronged ends with an oval shaped filtering member on each end. The filtering elements would be made of a medium grade cotton rolled into an appropriate configuration to fit the nostril, and could possibly be impregnated with oxymetazoline to aid in the treatment of nasal congestion. The flanged ends would secure the device in place during exhalation, and could also be useful in replacing the disposable filtering elements. The spring clip could be offered in clear or flesh-toned plastic and could also be offered in a variety of sizes to fit different users' noses.

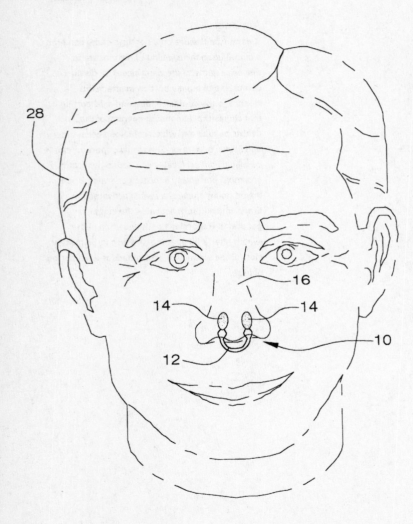

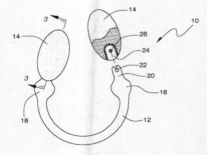

Patent No. D430,934
Sept 12, 2000
Charles E. Willard

Nose Pick

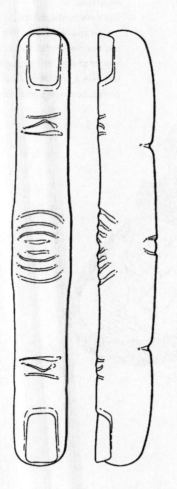

Abstract

The ornamental design for a nose pick, as shown and described. This is a plan view of a nose pick, showing my new design. This is a side view thereof with the opposite side view being a mirror image thereof. There is a bottom view thereof and an end view with the opposite end view being a mirror image thereof.

Patent No. D329,322
Sept 15, 1992
Jerry H. Wartell

Novelty Sandal

Abstract

The ornamental design for a novelty sandal, as shown and described. This is a perspective view of a novelty sandal, showing my new design, the slipper shown being intended for use on the left foot with a portion of a foot shown in dot dashed lines for environmental purposes only.

Novelty Tank Top

Patent No. US 6,434,751
Aug 20, 2002
Jamie K. Gustafson; Joyce A. Christy;
Stacy A. Kramer; Lynn E. Gustafson; Jody E. Howe

Abstract

A novelty tank top derived from a man's undergarment. The inverted undergarment is worn over the upper torso of the wearer. The novelty tank top has an anterior pocket for carrying small items and a spandex liner that improves the functionality of the tank top.

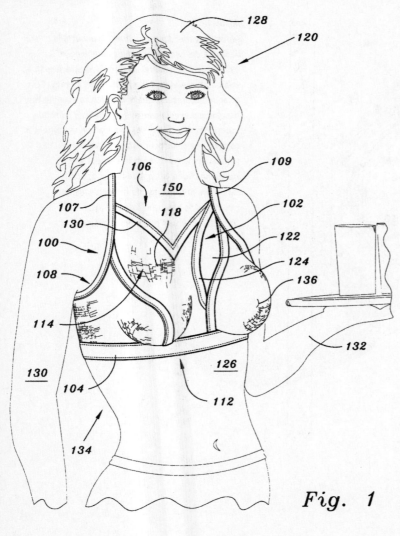

Fig. 1

Fig. 2

Fig. 3

Fig. 4

Patent No. US 6,655,061

Dec 2, 2003

Lawrence D. Good

One Way Free Spinning Hubcap

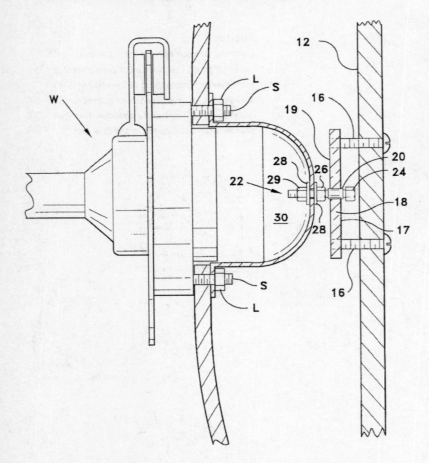

Abstract

A one way free spinning hubcap assembly having a decorative hubcap and a bracket assembly. The bracket assembly is attached to a vehicular wheel and the hubcap is rotatably attached thereto. A bearing clutch present in the bracket assembly permits the hubcap to rotate independently from the wheel when the wheel is rotated in one direction, but forces the hubcap to rotate concurrently with the wheel when rotational force is applied in the opposite direction. In this way the hubcap may be permitted to continue rotating after the associated wheel has ceased rotating.

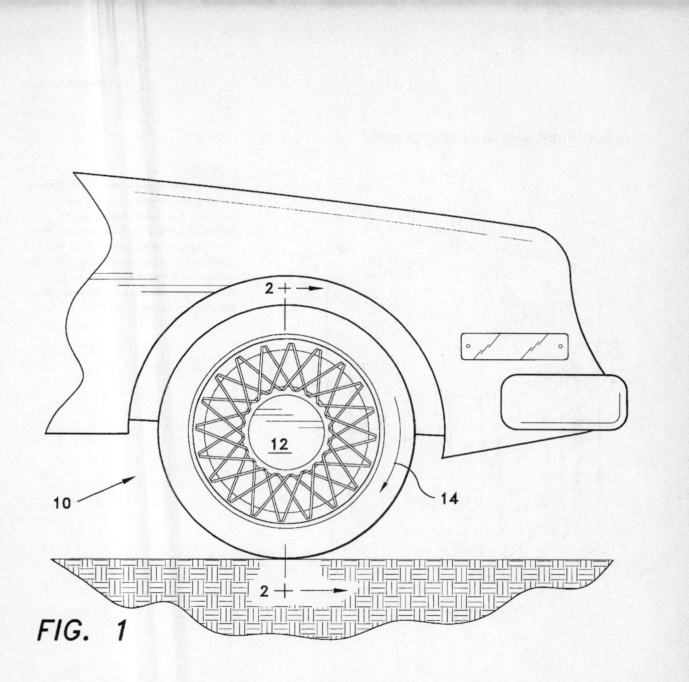

FIG. 1

Patent No. 3,806,723
April 23, 1974
Loren R. Ollom

Ornamental Display Arrangement

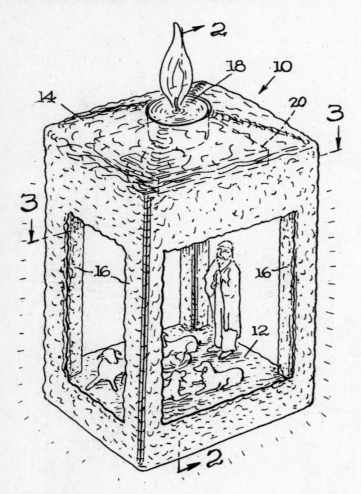

Abstract

An ornamental display arrangement comprising a kiosk or pavilion-like structure of wax or wax surface in which various figures may be disposed in a desired manner to depict a theme to be visually conveyed to an observer, said structure including a floor portion on which the various figures may be supported, said structure also including a roof portion on which illuminating means is supported.

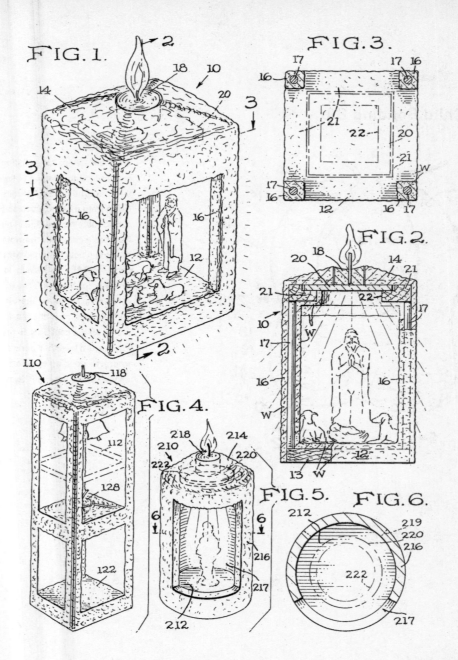

FIG. 1.

FIG. 3.

FIG. 2.

FIG. 4.

FIG. 5.

FIG. 6.

115

Patent No. 4,776,546
Oct 11, 1988
Alfred L. Goldson; Amy R. Goldson

Parent-Child Bonding Bib

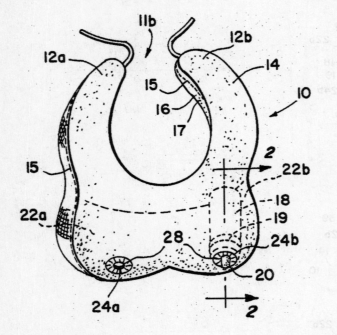

Abstract

A bib-like device for assisting in nursing a child
and improving bonding between a parent and
the child. The device includes a bib-like member
made from a fabric, such as terry cloth,
comfortable for the child. The device defined
includes a generally mammary-shaped area
defining a pouch therein for retaining a
container of liquid such as formula, milk, juice
or water for the child. The pouch is sized to
receive the container therein and communicates
with an opening in the fabric for permitting a
nipple on the container to protrude therefrom.
When worn by either parent, the device
improves bonding between the child and parent,
especially for the male, by anatomically
simulating the female.

FIG. 1.

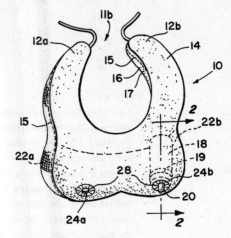

FIG. 2.

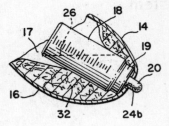

FIG. 3.

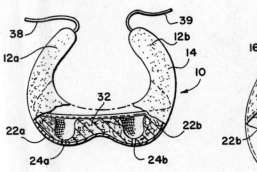

FIG. 4.

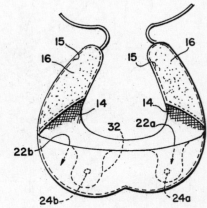

FIG. 5.

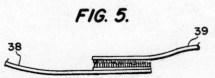

Patent No. US 6,325,069
Dec 4, 2001
Jules Heims

Pedicure System

Abstract

A pedicure system includes a post having a handle at the proximal end and a file at the distal end. A foot support has a plate with a lower end and an upper end. The upper end is formed with a downwardly projecting plate to retain the plate at an angle. A toe support is formed as a generally U-shaped channel with a lower horizontal portion adjustably secured to the plate and upwardly extending parallel side portions to allow the toe support to be shifted from side to side.

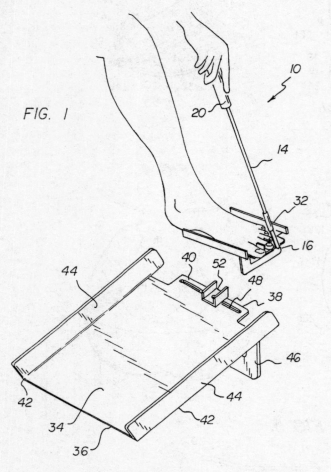

FIG. 1

FIG. 2

Patent No. 5,599,048
Feb 4, 1997
Thomas Schioler

Phosphorescent Book

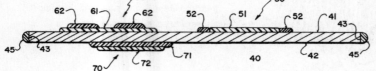

Abstract

A book is formed from a number of rigid pages, each formed from a stiff material. All of the indicia on the pages are formed from a material having luminescent properties and also having a thickness on the page to form a relief pattern on the page. The luminescent also includes a colored pigment providing different colors on the page. The indicium elements on each page are arranged relative to the next adjacent page so that they do not overlap when the pages are closed. An edge bead is applied onto relatively thick page edges to illuminate the edges and to provide a soft feel to the edges.

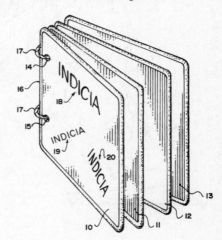

FIG. I

Patent No. US 6,575,808
June 10, 2003
Alvina L. Wright; Elana Burbank

Pom-Pom Puppet and Method of Cheering with Pom-Pom Puppet

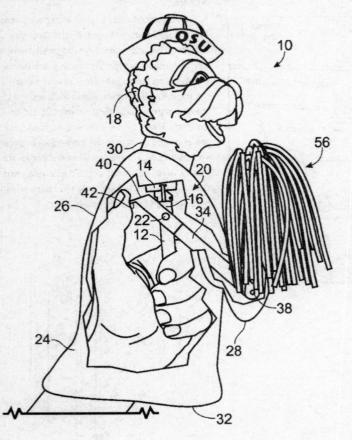

Abstract

A puppet for cheering having manipulable limbs having pom-pom attachments, limbs and a method of using the same. The puppet may include a handle portion that can be held by the user during use of the puppet, on which is attached one or more limb assemblies, and each limb assembly may include an upper limb element pivotally attached to the body support, an extending limb element pivotally carried on the upper limb element, a pom-pom attached to the extending limb element, a means to activate the limb assemblies, and a head connected to the handle portion. The method includes use of the puppet as a method of cheering, such that by manipulation of the limbs the pom-poms attached to the limbs move and the puppet appears to be the one cheering.

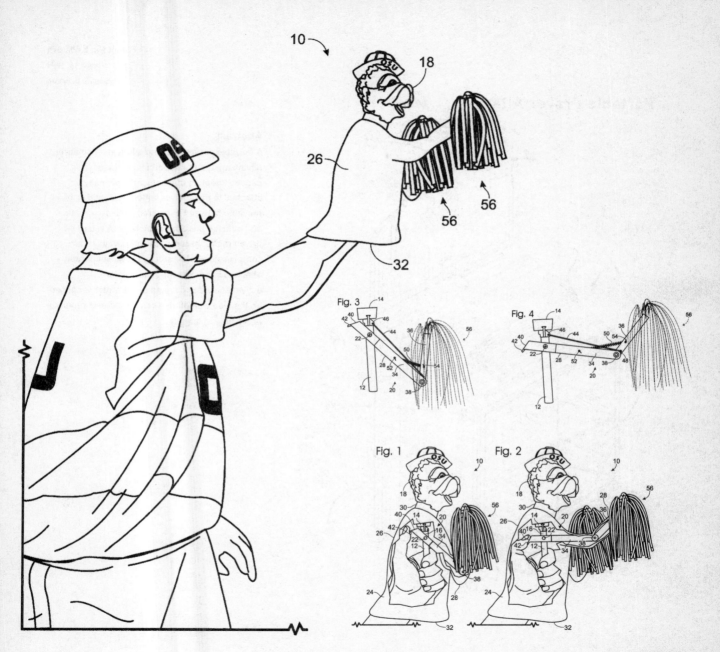

Fig. 3

Fig. 4

Fig. 1

Fig. 2

Patent No. 5,882,095
Mar 16, 1999
Donald E. Green

Portable Prayer Altar

Abstract

A portable prayer altar including an enclosure having an open rear wall facing the user. A padded kneeling mechanism is pivotally attached to the interior bottom wall and is movable between a lowered position for use, and a raised position for storage. A rotary air driven motor, in communication with an air compressor and air storage tank, is operably attached to the kneeling mechanism. An activation switch coupled to the motor is carried on the side wall of the enclosure and is readily accessible to the user.

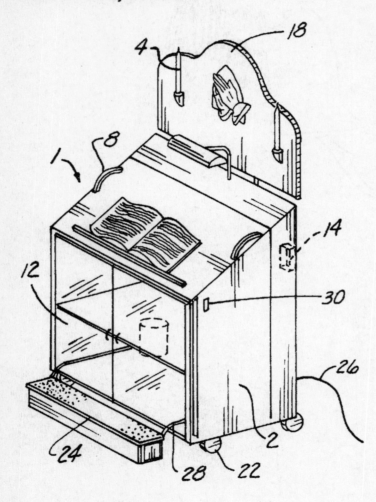

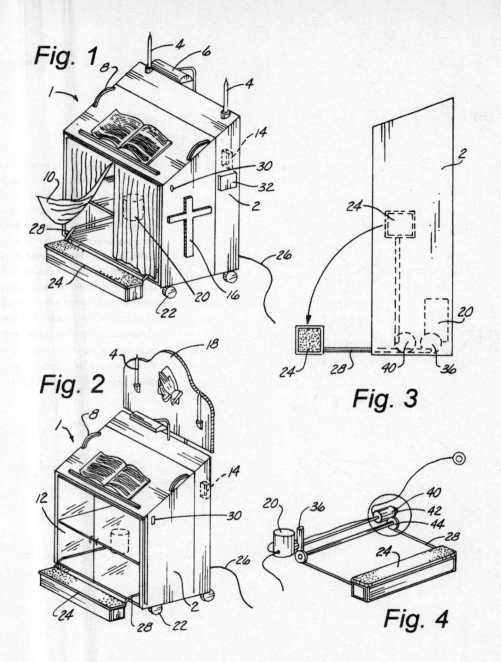

Fig. 1

Fig. 2

Fig. 3

Fig. 4

Patent No. 6,547,631
April 15, 2003
Suzanne L. Randall

Prayer Doll

Abstract

A prayer doll having a moveable head, limbs, and eyes, the moveable head, the limbs, and the eyes, pivotally mounted to the prayer doll, the other limbs, and the eyes; and a motion control system. The motion control system has: a motor and drive means, the motor driving the drive means, the drive means driving a plurality of cams, each of the cams driving a respective cam follower, each of the cam followers adjoined to a push pull cable at one end of the cable, each of the limbs, each of the eyes, and the head adjoined to a respective opposing end of the cable, each of the cable imparting motion to the respective head, limb, eye. The prayer doll may also have an audio playback system.

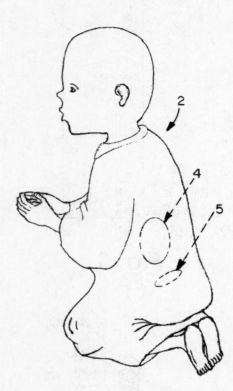

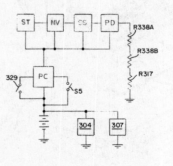

FIG. 14

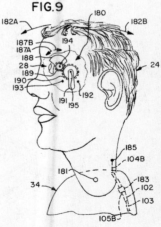

FIG.9

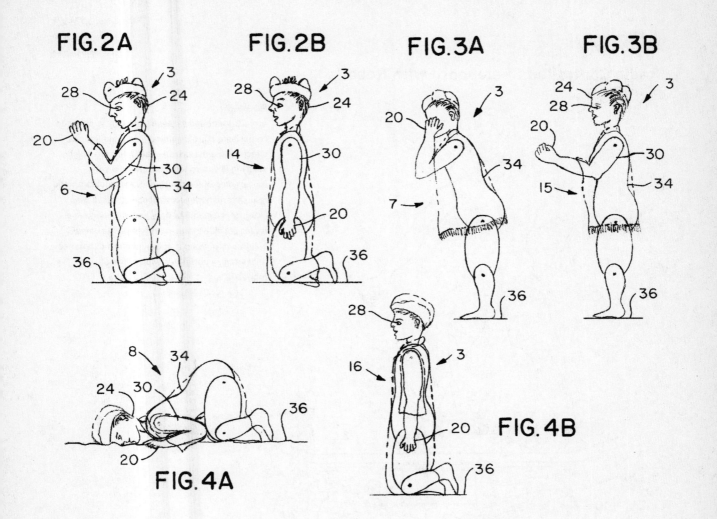

FIG.2A

FIG.2B

FIG.3A

FIG.3B

FIG.4A

FIG.4B

Patent No. 6,074,271
June 13, 2000
Steven Derrah

Radio Controlled Skateboard with Robot

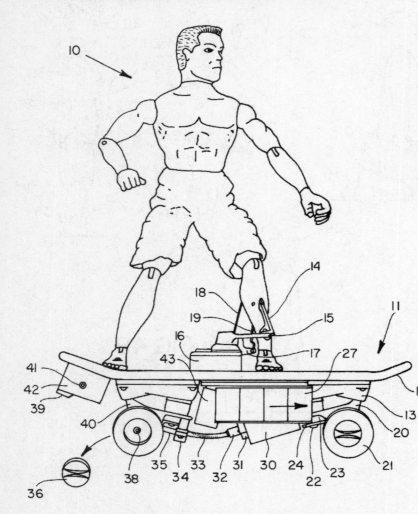

Abstract

A radio controlled skateboarding toy is provided comprising a multi-jointed moveable figurine attached to a motorized skateboard for which steering is controlled through the pivotal movements of the figurine allowing for dramatic realism and high performance stunts. It also features moveable battery packs, changeable motor positions, and interchangeable wheel weights to provide different centers of balance necessary to perform a wide range of maneuvers.

Patent No. D313,446

Jan 1, 1991

Andrew Froutzis

Religious Doll

Fig. 1

Fig. 2

Abstract

The ornamental design for a religious doll, as shown and described. This is a top perspective view of a religious doll showing my new design. Also a rear perspective view thereof, an enlarged front elevation view thereof, a left side elevation view thereof, with the right side elevation being substantially a mirror image.

Fig. 3   Fig. 4   Fig. 5

127

Patent No. D355,865

Feb 28, 1995

Quincy Stingley

Religious Holiday Basket

Abstract

The ornamental design for a religious holiday basket, as shown.

FIG. 1

FIG. 2

FIG. 3

FIG. 4

Patent No. 2,769,266
Nov 6, 1956
Robert S. Feeley

Religious Symbol

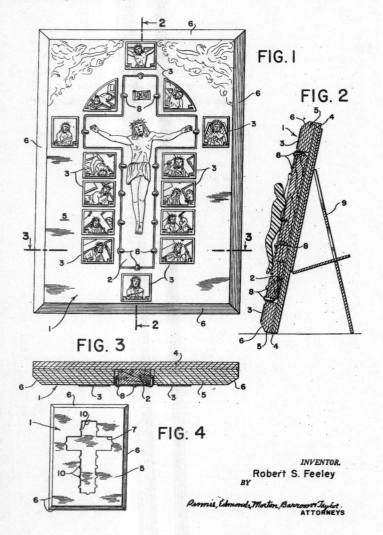

FIG. I

FIG. 2

FIG. 3

FIG. 4

INVENTOR.
Robert S. Feeley
BY

Pennie, Edmonds, Morton, Barrows & Taylor.
ATTORNEYS

Abstract

This invention relates to a religious symbol and provides an object especially adapted to use as a guide to prayer and particularly in the personal devotion to the Way of the Cross. In performing this devotion in a church, one usually moves about the church pausing at each station of the cross and there repeating the appropriate prayer. When unable to visit a church for this purpose, one may use a symbol of the respective stations, pausing at the respective symbols as though actually standing before the represented station. But, ordinarily, visual observation is necessary to distinguish the position of one station of the cross from that of another. By reason of blindness, or darkness or because one's eyes are closed in prayer, it is often difficult, if not impossible, to make this distinction visually. It is an object of my present invention to provide a symbol of the stations of the cross embodying means by which the position of the respective stations may be readily identified by the sense of touch during prayer.

Patent No. 3,701,395
Oct 31, 1972
Stuart J. Theobald

Rescue and Safety Vest

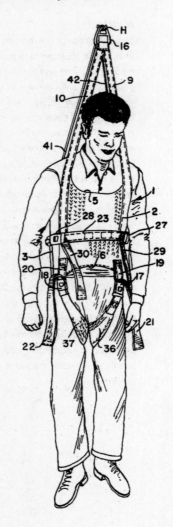

Abstract

A rescue and safety vest comprising separate front and back portions having means thereon for quickly and releasably joining the front and back portions together about the torso of a person. Both the front and back portions include enlarged vest portions which provide support and evenly distribute pressure over the torso of the person, and one of the portions included relatively wide leg straps which encircle the legs of the person and provide lifting support therefor.

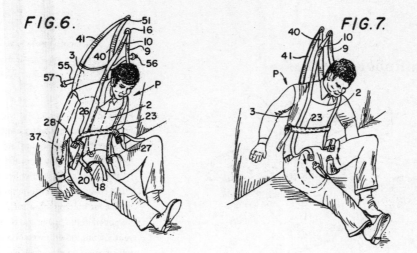

FIG. 6.

FIG. 7.

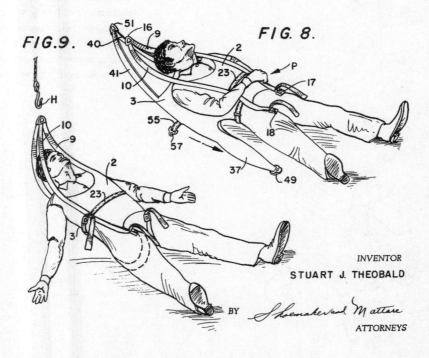

FIG. 9.

FIG. 8.

INVENTOR
STUART J. THEOBALD

BY *Shoemaker and Mattare*

ATTORNEYS

Robot System, Robot Device, and Its Cover

Patent No. US 6,505,098

Jan 7, 2003

Takayuki Sakamoto; Masahiro Fujita; Seiichi Takamura;
Yu Hirono; Hironari Hoshino; Nobuhiko Ohguchi

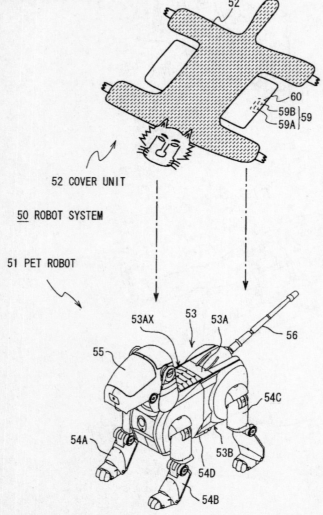

52 COVER UNIT

<u>50</u> ROBOT SYSTEM

51 PET ROBOT

Abstract

Firstly, an authenticating pattern is disposed on a cover and an authenticating device is disposed in a robot apparatus for authenticating the cover on the basis of the authenticating pattern of the fitted cover. Secondary, an information holding device for holding inherent information is disposed in the cover and a reading device for reading out the inherent information from the information holding device is disposed in the robot apparatus. Thirdly, a function of detecting an amount of influence due to the cover and changing manifesting patterns of motions as occasion demands on the basis of the detection result is disposed in the robot apparatus.

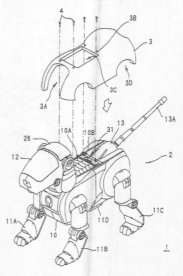

Patent No. US 2,882,858

Oct 15, 1956

Bertha A. Dlugi

Sanitary Appliance for Birds

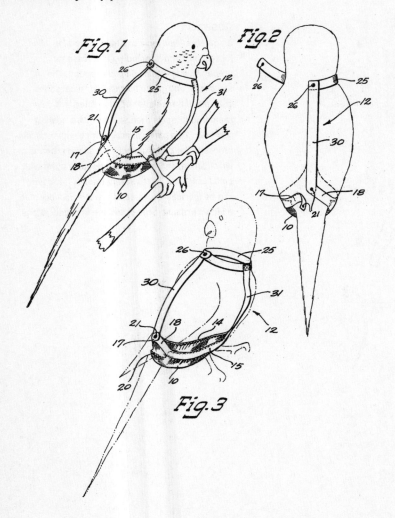

Fig. 1

Fig.2

Fig.3

Abstract

This invention relates generally to a garment to be worn by birds and more particularly to a garment having a patch of material especially adapted to be supported around the crissum of a bird for the purpose of receiving its excremental discharge.

Patent No. 5,196,240
Mar 23, 1993
Gregg M. Stockwell

Seamless Bodysuit and a Method for Fabricating Same

Abstract

A seamless bodysuit and the method for fabricating same, including a process for preparing a textile coating compound. The seamless, one-piece bodysuit for a person consists of a textile foundation fitted to a mannequin that is then sealed with a coating compound prepared in accordance with a novel process. The bodysuit includes sealing means at wrist, ankle, and neck openings for mating with suitable gloves, boots and helmet or hood and can be fabricated for use as a wet suit, a dry suit, a biohazard suit, or in other similar applications.

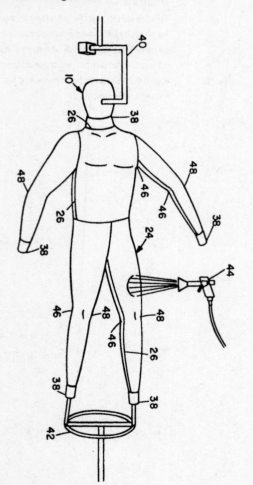

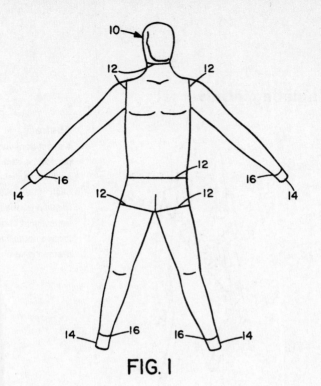

FIG. 1

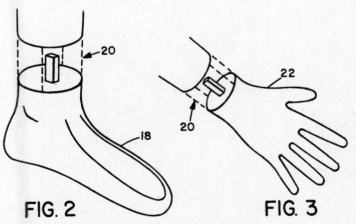

FIG. 2

FIG. 3

Patent No. US 6,659,840

Dec 9, 2003

Perry R. Chastain; Stacey T. Henderson

Set of Dolls for Simulating a Universal Beauty Pageant

Abstract

A set of dolls for simulating a universal beauty pageant, including a plurality of female dolls, each being distinguishable from one another. Each of the dolls is wearing a dress with an identifying ribbon thereon. A plurality of educational cards correspond with each of the female dolls. Each of the cards have information thereon relating to a corresponding female doll.

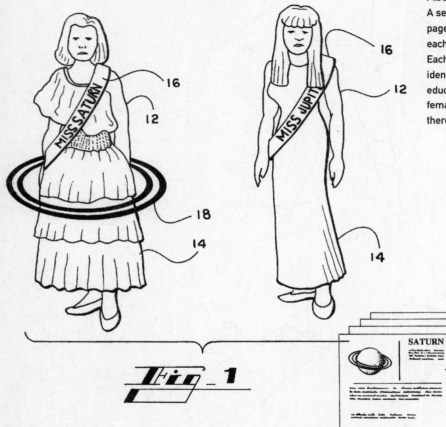

136

Patent No. 4,089,067
May 16, 1978
Rebecca L. Velasco

Shirt

Abstract

A shirt having a tubular body portion whose axis would be substantially vertically oriented when being worn by a person. The tubular body portion has a front section and a back section. It also has a top, middle and bottom area. At least four arm sections extend outwardly from the tubular body portion somewhere between its top and middle area and at least two of these arm sections extend from the back section. The tubular body portion has a body aperture formed in the lower edge of the bottom area and has only one neck aperture formed in the top edge of the top area.

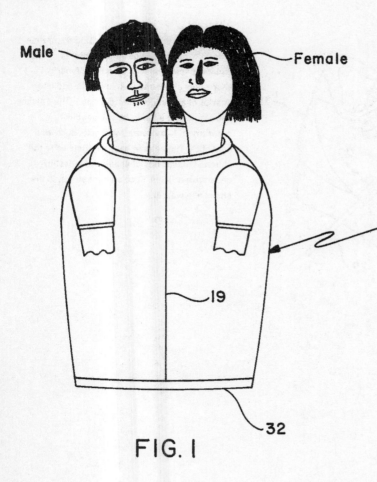

FIG. 1

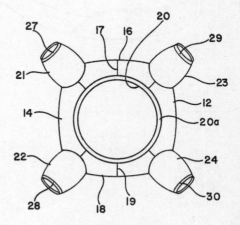

FIG. 2

Patent No. 5,584,132
Dec 17, 1996
Henry Weaver; Brent Weaver; Ronald Perryman

Shoelace Tip Holder

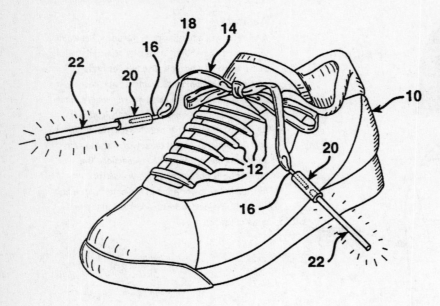

Abstract

An adapter is provided for attaching decorative articles to shoelace tips. The adapter is small and inconspicuous with an outer cylindrical shape only about one inch in length and one quarter of an inch in outer diameter. The adapter has at one end a fitting for releasable securement to a decorative article, such as a light stick. The adapter may be attached to the stiffened plastic tips of shoelaces to fasten decorative articles, such as light sticks, to the shoes of a wearer.

FIG. 1

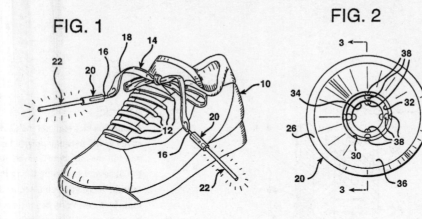

FIG. 2

FIG. 3

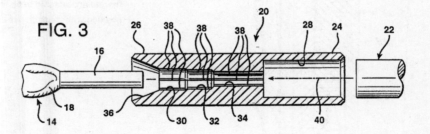

FIG. 4

FIG. 5

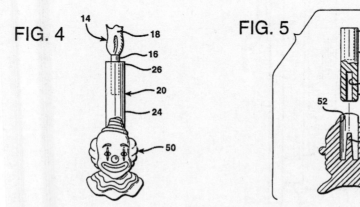

Patent No. 3,392,821
July 16, 1968
William J. Tracey

Sick-Call Set

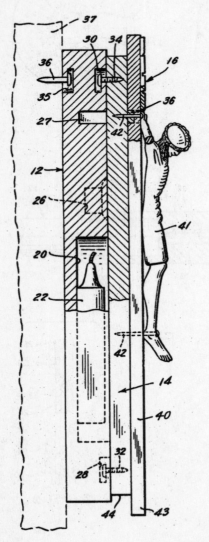

Abstract

A religious sick-call set including a base, in a cavity of which religious articles are contained and further including a cover plate for overlying the base and enclosing the religious articles therein, a crucifix being secured to the cover plate and extending beyond an end thereof for being received in a socket formed in the base, wherein the crucifix and the cover plate secured thereto are adapted to be mounted in a vertical position with respect to the base.

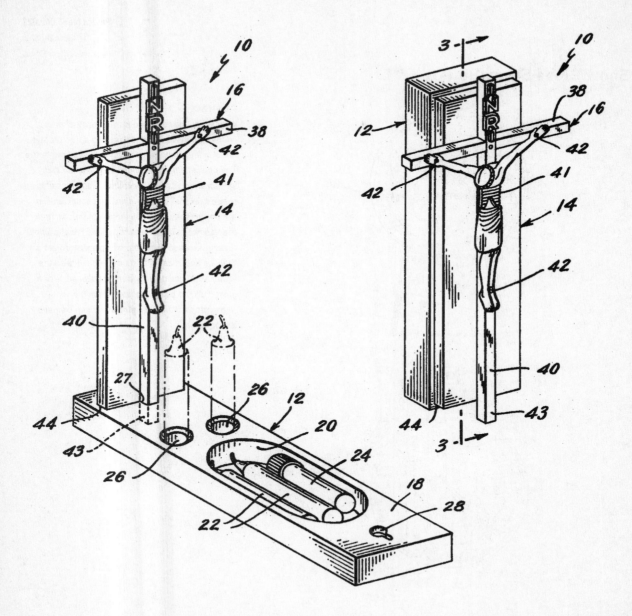

Patent No. 6,042,022
Mar 28, 2000
Nancy E. Rogozinski; Larry Couey

Snow Globe Spray Bottle

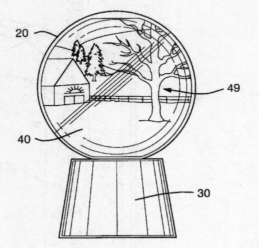

Abstract

A spherical bottle has a pump spray atomizer connected at a neck opening and a thin plastic sheet with printing or a graphic inside the bottle. The bottle contains a liquid fragrance or body splash which can be atomized. Flakes or chips resembling falling snow are dispersed within the bottle. The chips are large enough so that they do not fit into the spray pump mechanism. A ceramic base receives the spray atomizer so that the bottle can stand on the base with the spray atomizer down when it is stored. When stored on the base, the bottle has the appearance of a snow globe and can be used decoratively.

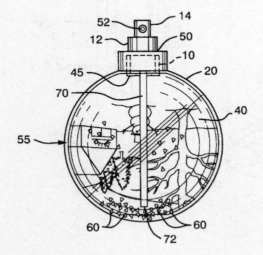

Patent No. US 6,199,216
Mar 13, 2001
Rose M. Weatherspoon

Sock with Pocket

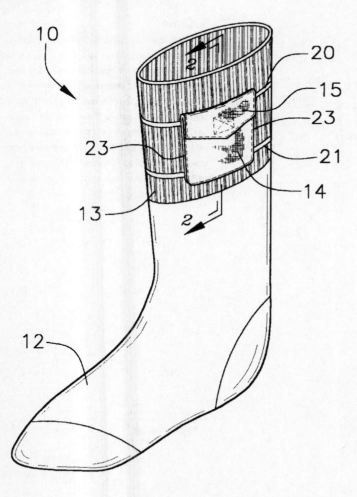

Abstract

A sock with pocket for storing items. The sock with pocket includes a foot portion and a leg portion extending from the foot portion. A pocket flap is coupled to the leg portion and forms a pocket between the flap and the leg portion. A cover flap is coupled to the leg portion and folds over an upper edge of the pocket flap for closing the pocket. The cover flap is detachably couplable to the pocket flap.

Patent No. 6,030,271
Feb 29, 2000
Michael Pietrafesa

Soft Baby Doll

Abstract

A baby doll whose torso is defined by a fabric bag having a pair of arm openings, a pair of leg openings and a neck opening. Coupled to the arm openings and extending therefrom are hollow, flexible-plastic arms, coupled to the leg openings and extending therefrom are hollow flexible-plastic legs, while mounted above the neck opening is a hollow, flexible-plastic head having a neck coupled to the neck opening whereby the arms, legs and head are articulated from the torso bag.

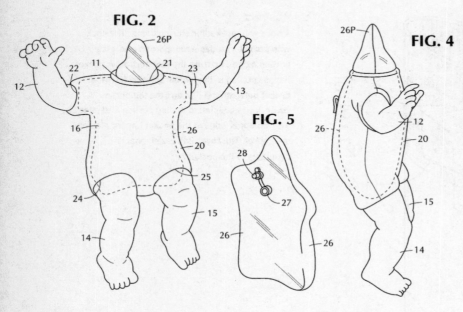

FIG. 2

FIG. 5

FIG. 4

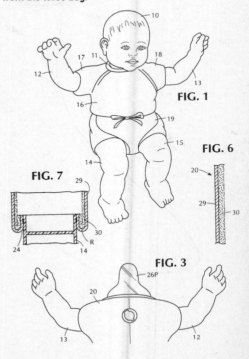

FIG. 1

FIG. 6

FIG. 7

FIG. 3

144

Patent No. 5,484,316
Jan 16, 1996
Mark A. Poirier

Sound Effects Belt

Abstract

A novelty belt with a control pad for programming
and actuating a variety of sound effects emitted
through a speaker. The belt also includes a
plurality of props that correspond to the sound
effects, creating a thematic novelty toy.

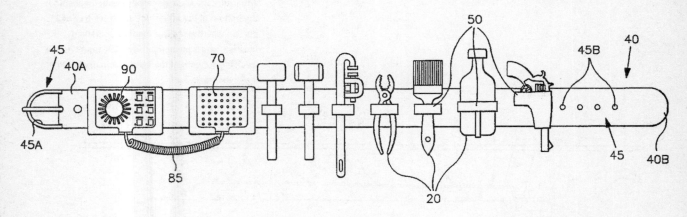

Patent No. 6,145,442
Nov 14, 2000
Peter Anthony Lofts

Submarine Amusement Ride

Abstract

An underwater mobile observatory system comprising an aquarium able to hold water and large enough to support fish, coral, and to display artificial objects such as shipwrecks and ruins, a vehicle track extending through the aquarium, the track generally being adjacent to the bottom of the aquarium, the track having a portion which rises to a loading/unloading position, and a passenger vehicle coupled to the track for movement therealong and unable to leave the track.

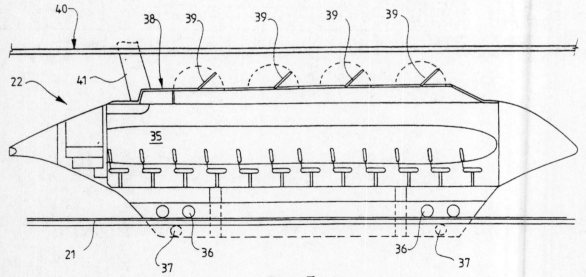

Fig. 5

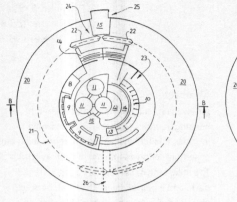

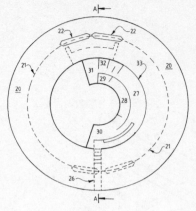

Fig. 1

Fig. 2

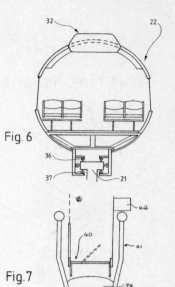

Fig. 6

Fig. 7

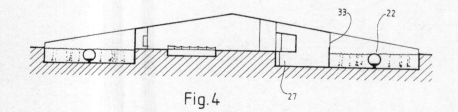

Fig. 3

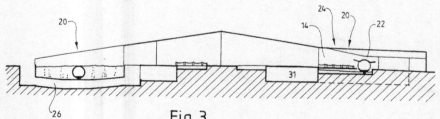

Fig. 4

Patent No. 5,983,411

Nov 16, 1999

Herbert Demoret

Toilet Tank Aquarium

Abstract

A new toilet tank assembly aquarium for housing aquatic creatures. The assembly includes a toilet bowl assembly with a toilet tank assembly coupled to the toilet bowl assembly. The toilet tank assembly has a top, a front, a back, and a pair of sides. The front, back and sides of the toilet tank assembly define an interior space with the top of the toilet tank assembly having an opening into the interior space of the toilet tank assembly. A toilet water reservoir is located in the interior space. A lid substantially covers the opening of the top of the toilet tank assembly. The front of the toilet tank assembly is generally transparent.

Patent No. US 6,688,040

Feb 10, 2004

Rong Teai Yang

Tombstone Flower Saddle

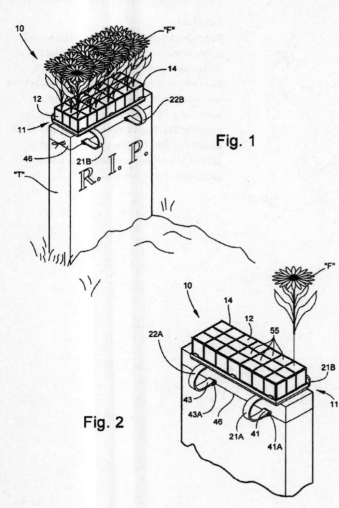

Fig. 1

Fig. 2

Abstract

A flower saddle includes a base adapted for securing the flower saddle to a supporting structure. A foam block is located on the base, and is adapted for receiving stems of flowers and for holding the flowers in a desired position for display. A cage includes openwork covering the foam block. The openwork defines access points through which the stems of the flowers are received into the foam block.

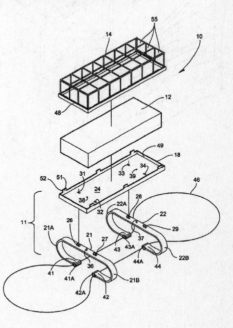

Patent No. 6,655,822
Dec 2, 2003
Ty M. Sylvester

Trailer Hitch Cover

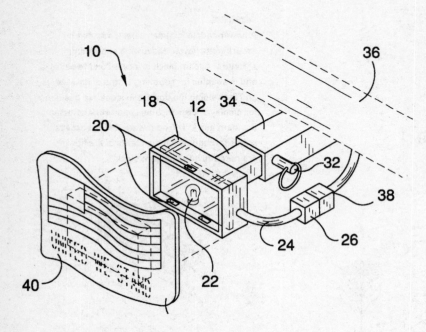

Abstract

Decorative trailer hitch covers protect and block a vehicle trailer hitching socket from view in a decorative manner. Decorative trailer hitch covers prevent water, dirt, and other debris from accumulating in the vehicle trailer hitching socket when it is not in use. Furthermore, the normally unattractive trailer hitching socket can be covered with an attractive, illuminated graphic or sign. In this manner, the driver can personalize his or her vehicle.

Patent No. 5,149,289
Sept 22, 1992
Patricia Edwards; Jody Armstrong

Transformable Doll

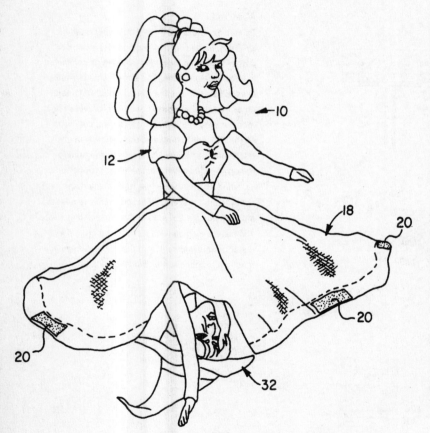

Abstract

The invention includes a doll transformation between two characters or persona. The doll includes interconnected upper body portions of each character or persona. A stand portraying the lower body portion of at least one of the characters or persona receives the upper body portion not in use. Assembly of the stand with one of the upper body portions extending therefrom gives the illusion of a complete character or persona. A number of characters portrayed by dolls are transformed between two, and sometimes more, personalities or likenesses in their story-book or fantasy lives. For example, Cinderella's rag clothes are instantly transformed into a beautiful gown by her fairy Godmother, the evil queen disguises herself as an old woman in order to trick Snow White, and the Little Mermaid who was transformed into a princess by the wicked Sea Witch. Unfortunately, dolls previously available have not been capable of recreating such transformations. Typically, dolls require a complete change of costume, a task very unlike the magical change often undergone in the character's fantasy story.

Patent No. 5,309,651
May 10, 1994
David B. Handel

Transformable Shoe

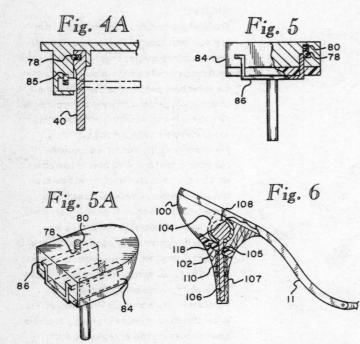

Fig. 4A

78
85
40

Fig. 5

84
80
78
86

Fig. 5A

80
78
86
84

Fig. 6

100
104
118
102
110
106
108
105
107
11

Abstract

Transformable shoe frames are described wherein the heel of the shoe can be extended downwardly in a high heel position or converted to a low heel position. The shoe frame is made in sections that are pivotally connected to each other and mechanisms have been provided to lock the sections together in varying positions, so that the angle between the toe portion and the remainder of the shoe can be adjusted to maximize the comfort of the wearer depending on whether the heel is in a high heel or low heel position. In a preferred embodiment the heel is stowable under the sole of the shoe and the sole will have a well defined flexible region in the distal metatarsal region of the shoe, either by having a hinged joint or built in flexibility.

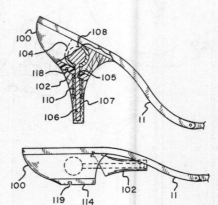

100
104
118
102
110
106
108
105
107
11

100
119
114
102
11

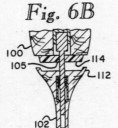

Fig. 6B

100
105
102
114
112

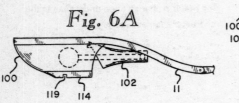

Fig. 6A

100
119
114
102
11

Patent No. US 6,227,216
May 8, 2001
Isaac B. Dweck

Umbrella Having Ears

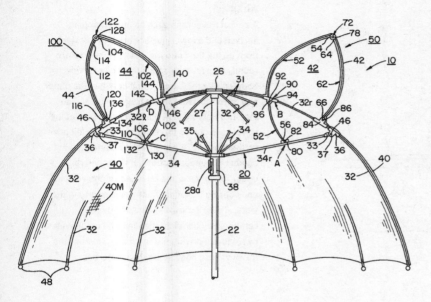

Abstract

A collapsible umbrella is provided having ear-like projections as part of its main canopy, and includes an umbrella frame with a central shaft having a handle at one end and a crown member at the other end. The umbrella frame further includes a first ear-like frame assembly and a second ear-like frame assembly, with the main canopy including first and second ear-like covers that extend over the first and second ear-like frame assemblies, respectively.

Patent No. 5,775,226
July 7, 1998
Hiroshi Futami; Kenjiro Futami

Underwater and Land Travel Vehicle

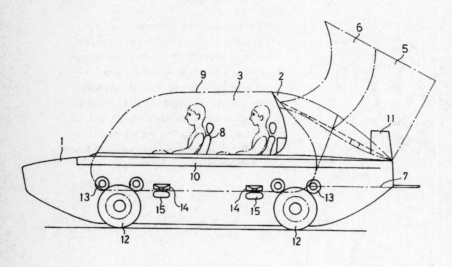

Abstract

A pair of travel rails each having an upper surface and a lower surface are laid to extend from on the land into the water. Main tires are rotably disposed on opposite sides of a vehicle body, so that they are located between the upper and lower surfaces of the travel rails, and so that they abut against the lower surfaces of the travel rails during traveling of the vehicle body on the land. Auxiliary tires are adapted to abut against the upper portions of the travel rails under influence of the buoyancy of the vehicle body during traveling of the vehicle body in the water. Thus, the travel vehicle body can continuously travel on the land and in the water by rotatively driving the main tires to move the travel vehicle body along the travel rails.

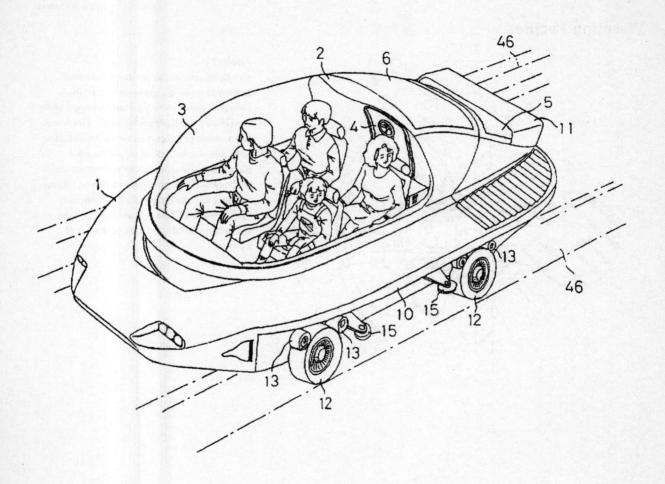

Patent No. US 6,264,678
July 24, 2001
Sonny B. Landers

Vibrating Pacifier

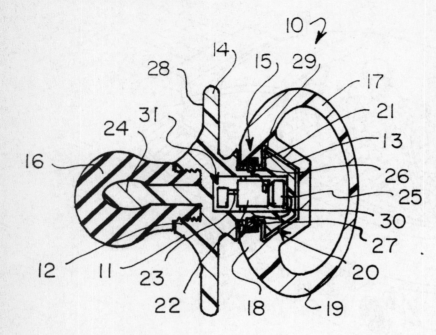

Abstract

A vibrating pacifier for providing increased comfort to teething child in particular. The vibrating pacifier includes a body having a cavity therein and also having a front end, a back end, and a mouth guard portion extending about the body; and also includes a nipple member disposed at and extending from the front end of the body; and further includes a handle member rotatably and sealingly mounted to the body near the back end thereof; and also includes a vibrating assembly for vibrating the nipple member.

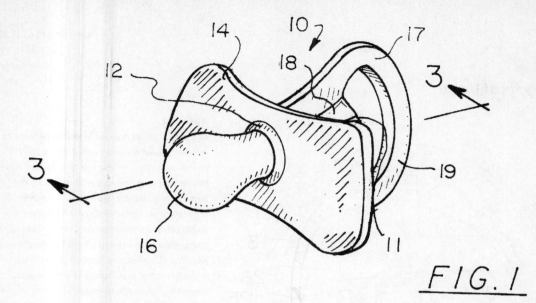

FIG.1

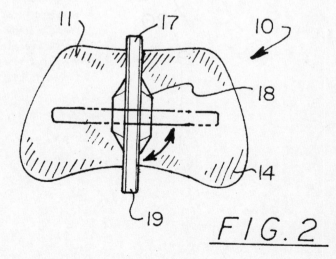

FIG.2

157

Patent No. US 6,401,260
June 11, 2002
Timothy Porth

Wobbling Headpiece

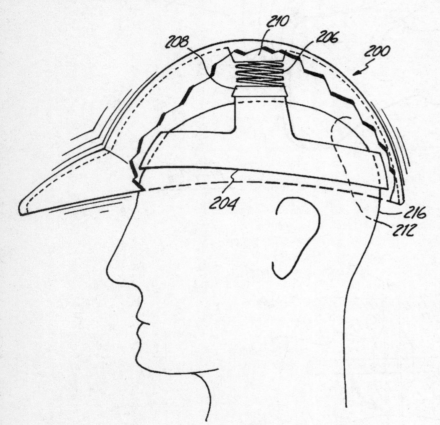

Abstract

One aspect of the present invention pertains to a wobbling headpiece that includes a display member having an inner concave portion that substantially surrounds and is substantially disassociated from a head strap. An action mechanism is operably disposed between the display member and the head strap. Various headpieces have been designed to attract the attention of bystanders through the incorporation of mechanical and/or electrical elements. For example, headpieces known in the art incorporate battery-operated fans, flashing and/or rotating lights, and incorporating rotating ornaments. At least one headpiece known in the art incorporates a dangling element designed to dangle mistletoe over the head of a wearer. In the view of the foregoing, there is an ongoing need for unique headpieces that appeal to the public-at-large and draw attention to those who wear the headpieces.

SEND US YOUR BRILLIANT IDEAS!

Have an idea that you think is legendary? Epic? Massive? Brilliant? Insane?

Go to our Web site at www.patentlyridiculous.com.

If you are one of the lucky people whose idea is accepted, you will receive a copy of the book and a certificate designating you as a contributor of a ridiculous (although possibly brilliant) invention.

Submission requirements:
Send a one-page or less description of your idea. If you want to send an illustration, send a scan as a tiff image. This is not necessary, but welcomed. If your idea is accepted, our mechanical engineers will translate your idea into a visual image. Include your name, mailing address, telephone number, and age, and the title of the invention.

By submitting, you are NOT being granted a patent. This site and book are NOT agents of the United States Patent Office. By sending your idea, you are granting permission to the author, his assignees, and the publisher to use the idea for publication and any manifestations associated with a book tentatively titled *Patently People's Ridiculous*, or variations of such name.

If you would like to have your idea patented, there will be directions to the proper U.S. government offices on the Web site www.patentlyridiculous.com.

Want to purchase any of the items pictured in this book? See what has been brought to market by visiting the site www.patentlyridiculous.com.

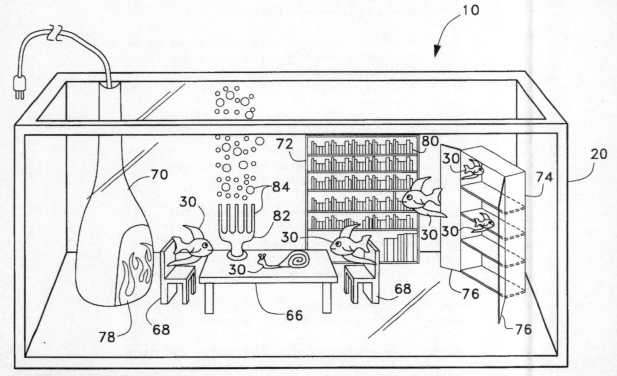

Combination Aquarium and Furniture System